Abbeydale Park Rise
Photograph: Ian Wilshaw

Persons Unknown
The Battle for Sheffield's Street Trees

Simon Crump and Calvin Payne,
with Julie Stribley

Foreword by Nick Hayes, author of *The Book of Trespass*

In memory of John Errington, Mary Marshall and Roy Millington.

About the authors

Calvin and Simon both received suspended prison sentences for breaching a High Court injunction, brought by Sheffield City Council, to prevent peaceful direct action by protesters seeking to stop the felling of healthy street trees.

Calvin is a community activist and local historian. Simon is a novelist and short-story writer. Julie is a member of Sheffield Tree Action Groups and a full-time nature lover.

Persons Unknown
The Battle for Sheffield's Street Trees

Simon Crump and Calvin Payne with Julie Stribley

All rights reserved. No part of this book may be reproduced, stored in a retrieval system or transmitted in any form or by any means electronic, mechanical, photocopying, recording or otherwise, without the prior permission of the publisher.

ISBN 978-1903110881

First published in this edition 2022 by Wrecking Ball Press

Copyright © Simon Crump, Calvin Payne & Julie Stribley

Book design: humandesign.co.uk

Copy editor: Sheldon Hall

All rights reserved.

This is a work of nonfiction. No names have been changed, no characters invented, no events fabricated. The opinions expressed in this publication are those of the authors and those interviewed who have provided anecdotes. The views, information, and opinions expressed are solely those of the individuals involved and do not reflect or represent the opinions or views of Wrecking Ball Press and its team.

CONTENTS

Foreword	11
Introduction	15
Preface	17
Wayland Road	19
Bannerdale Road	25
Marden Road	31
Rustlings Road	39
Raven Road	47
Chippinghouse Road	53
St. Ronan's Road	71
The Injunction	79
Olive Grove Road	83
Woodstock Road	89
Sheldon Road	93
Dunkeld Road	99
Kenwood Road	107
Edgedale Road	119
Lismore Road	127
Spring Hill Road	137
Vainor Road	147
Rivelin Valley Road	153
Vernon Road	169
Chatsworth Road	173
Millhouses Lane	177
Abbeydale Park Rise	185
Coverdale Road	201
Thornsett Road	209
Meersbrook Park Road	221
Tales From The Flying Squad	245
The Real 'Persons Unknown'	267
Afterword	291
Principal Persons: AKA The Opposition	295
Glossary	297
The Legal Stuff - Court Reports	301
Author's Note	359
Diary of a Tree Protester	361
Acknowledgements	387

FOREWORD

It was a telephone call from a former teacher of mine that first got me entangled with the Sheffield tree campaign. Rob Macfarlane, the Indiana Jones of English nature writing, had called to see if I could illustrate a poem he was writing in aid of the Sheffield tree campaign. He called the poem a 'charm against harm' but at that point neither of us knew exactly how it would be used.

The poem was called *Heartwood*. An ode to the life of a tree, and the life it sustains, it speaks from the tree's perspective, directly addressing the nameless worker who had come to cut it down. The poem evokes the universe of each tree, 'a city of butterflies, a country of creatures', and conjures a memory of what trees have always been in our culture, gods that watch over the scurrying of humans with a time-drawn wisdom. The final lines reminded me of another great poet of the pastoral, John Clare, and his poem 'Lamentations of Round Oak Waters'. Here Clare also laments the loss of the trees, but this time as symbols of the commons, areas of land to which local people not only had rights, but also exercised a responsibility of management. The commons were never just places, but an architecture of shared power, a reciprocal relationship between community and environment, that sustained nature and society for generations to come. As the commons were enclosed, so this reciprocal relationship became eroded, subjugated to a new order. Key to Macfarlane's and Clare's poems is the idea that there is a wider picture to be considered – it is not the tree cutter who is to blame, but the system that legitimises their actions. In Clare's words:

Although their aching hands did wield
The axe that gave the blow
It was not them that owned this field
Nor planned its overthrow.

John Clare was writing in the 19th century, a period of intense enclosure where laws were created by an unreformed parliament (i.e., powerful landowners) to enforce their privatisation of common land. When commoners lost their rights to the land, the right to self-subsistence, warmth and shelter, they also lost their ability to guard it from exploitation. And those who know the English history of enclosure also know that it hasn't ended. It has clawed its way from the countryside, where it began, into the cities, where not only public spaces have been privatised, but also public amenities. Council-run services have been contracted out to private firms, whose legal obligations are to their shareholders, and not the citizens of a city.

When the residents of Sheffield stood up against the felling of their street trees, they were fighting a modern enclosure, the Private Finance Initiatives (PFIs) that took democratic decision-making from the people into the sphere of privatised cost-cutting. They began as isolated groups, neighbours trying to defend the trees on their particular road, but it was when these groups joined together, to form STAG, that the campaign really gained its momentum. By standing in between Private Finance Initiatives and the trees, they were taking back their rights to defend the nature they loved so much – they were reclaiming the commons.

And what they faced was a quick-fire synecdoche of the entire working-class history of the land. By-laws were created, injunctions were served, fences were erected around trees, interdicted with special spells that could land protesters in prison if breached. They faced slur campaigns, they faced losing their liberty, and since many were educators, they risked their jobs and livelihoods. But the residents of Sheffield wanted their trees, they wanted their birdsong, they wanted the green. They were courageous, they persisted, and they won.

The poem *Heartwood* and its illustration were used to raise money to pay for legal costs for some protesters. The legal costs alone were an example of how faceless power, private firms with lawyers on retainer, operate on those without money – the costs were a punishment for those who stepped against the power, and a deterrent for those to come. But suddenly, with Rob's poem, his talent and sway, the campaigners were able to raise enough money to pay legal costs, and keep the protesters out of prison. I watched in awe as the protesters sold prints of the poem, publicising both the sales and the cause, using Twitter to garner support from celebrities and Government officials alike. In spite of illustration being my chosen career, I had always seen the arts as something decorative, the icing on the cake of society. But here, words forged inside a poet's head, painted in ink by an illustrator, were dissolving the walls of a prison, permeating the power of a system, being used to keep people free. Rob was right. This really was a charm against harm. It was something like magic.

But the Sheffield tree campaigners taught me the true substance of this magic. Too often the media represents grassroots campaigns as doomed to fail, and when they succeed, claim it had nothing to do with the campaign. Too often protesters are objectified in the words of our Prime Minister as 'un-cooperative crusties', dismissed as idle layabouts, utopian hippies, woolly-headed dreamers, too naive to the workings of power. But the Sheffield tree campaigners know differently. You don't need a bullring through your septum to protest inequality. You don't need green hair to stand up to power. You need to collectivise. They have taught us that the key to challenging power is connectivity, the WhatsApp groups that alerted each other to a morning's fellings, the solidarity between protesters keeping each other out of prison, the tireless emails, conversations and networks created to stand up to the cold, blind logic of a PFI contract.

The *Heartwood* poem is still selling. The protesters have long been acquitted, so now the funds raised are being used to finance Lawyers for Nature, another grassroots organisation set up to evolve environmental law, to give trees and rivers legal rights in themselves, what the lawyers call 'standing'. The legacy of the Sheffield tree campaign continues to this day: in the hearts of protesters in muddy camps around England, who have proof that protest really works, standing up against the stark illogic of HS2 and in the trees themselves, still greening the streets of Sheffield, still standing.

Nick Hayes

Author, *The Book of Trespass*

INTRODUCTION

Persons Unknown means you. In High Court injunction terms, a 'Person Unknown' is anyone not named in an injunction, but instructed by a judge not to do something. You, yes I mean you, faced committal to prison, fines, and crippling legal costs if you tried to prevent a tree being felled in Sheffield between August 2017 and January 2020. There are similar injunctions enabling environmental damage across this country. You are a 'Person Unknown'.

Herein are first-hand testimonies and stories of people in Sheffield (and a bloke on a bike from Barnsley) who took direct action to protect the environment or maybe just 'their' tree, or all our trees on our streets. All defied their council; some their neighbours and friends. Some defied the High Court.

Reading the individual accounts, collated over two years of interviews, it is striking how little of what happened was planned in advance. These are stories of how people reacted to what was happening in front of them. Everyone who has contributed to this book has at one time stood on the street in chainsaw dust and chippings as a hung, drawn and quartered tree is carted off on a lorry. None wanted to see it again. They chose to do something, anything they could, and found themselves crossing lines and taking risks they'd previously thought impossible for them.

I had the opportunity to speak to a campaigner against fracking from the camp at Preston New Road when they attended a small 'training' event in Sheffield in 2017. 'You guys are amazing', he said. 'We have a camp, we know when they are coming and where they have to get in. You have to protect trees all over the place'. But we had people all over the place, people standing on their street corners or twitching curtains, standing on step ladders looking into the Council depot and chasing felling crews in cars. These are the stories of people prepared to step over a barrier and refuse to move.

At the heart of direct action were Calvin and Simon. Two people for whom the committee-determined, planned, strategic organisation and 'back room' campaigning were often an anathema. Why? Because three years of rigorously researched, expert-evidence-based challenges to Sheffield City Council's secret contract and flawed 'tree replacement' scheme had failed to stop the wanton destruction of healthy trees and the deliberate misleading of the public. Massive, healthy, mature trees were coming down.

The stories Simon and Calvin have collated, and contributed to, capture the despair, stupidity, heartbreak, absurdity and wilful disrespect for authority that comes from people 'having a go'. They stood against the might of a secretive council, a clueless police force, a legal system designed to crush protest under the weight of financial threat, and a corporate multinational bent on completing a lucrative contract that was supposed to be about better roads.

What of the trees? Well, a lot were felled, but far more were protected. A lot of people learned the value of peaceful, non-violent direct action when you just know, without a doubt, that something is wrong. Their stories will live on.

They did have the advantage of chainsaws, grabber trucks, Heras fences; they had Inspector Stubbs with a legion of police; they got High Court Injunctions, brought in private security guards, and had some highly motivated tree fellers. But we were right. We were smarter, more determined, more creative, and downright funnier. And, in the case of Simon, better dressed. And we had Calvin, to whose stubborn determination we owe so much.

Paul Brooke

Former Co-Chair, Sheffield Tree Action Groups (STAG)

PREFACE

'The story of ordinary people doing extraordinary things in the service of their community.'

In 2012 Sheffield Labour Council and the Department of Transport signed a twenty-five-year contract with Amey PLC costing £2.2 billion of public money to renew the city's highways in a programme titled 'Streets Ahead'. That contract has never been made fully publicly available. As a result of persistent Freedom of Information requests, we now know that it includes the following clause: Amey 'shall replace the highways trees in accordance with the annual tree management programme at a rate of not less than 200 per year so that 17,500 highway trees are replaced by the end of the term'.

While a small number of these trees were genuinely dangerous, dead, diseased, or dying, the Council introduced two further categories of trees to be removed: 'damaging' and 'discriminatory'. It was these last two of the '6 D's' that caused the controversy. Trees were blamed for damage to the roads and pavements and the Council claimed that they prevented people passing by in wheelchairs or when pushing prams.

The early days of the campaign concentrated upon engaging expert opinion, informing the public, challenging the Council's arguments and policy, and presenting them with an alternative. This was done successfully, and as a result a large and well-informed campaign began. However, it soon became clear to many activists that something else needed to be done. From 2015, until the tree-felling programme was 'paused' in March 2018, people in Sheffield carried out 'non-violent direct action' (NVDA) to prevent the unnecessary felling of healthy street trees. This book isn't about the issue as a whole, it isn't a history of the campaign, or of Sheffield Tree Action Groups (STAG). What it is about is telling the story of ordinary people doing extraordinary things in the service of their community.

While some of the stories told here are well known in the campaign, many are not. Some are of mass events, many are of individual actions. All the chapters consist of original first-hand accounts of events from the perspective of the central characters.

There are many other important parts of the campaign, not least the memorial trees on Western Road in Crookes or the 'Chelsea Elm' on Chelsea Road in Nether Edge, which, fortunately, didn't see any attempts to fell them. That they didn't was down to the strength of the organisation and local campaigns in those areas.

This book is intended as a resource; what is presented here is data, not data analysis. The protesters speak for themselves, and often the stark truths they speak are all the more shocking for that.

June 2015

'You're not having this fucking tree!'

Nicky Cowan, Resident

Wayland Road is a residential street in the Sharrow Vale district of Sheffield. It is lined with small mature cherry trees.

Nicky Cowan: In 2015 I began to attend public meetings at the Town Hall about what Sheffield City Council described as 'tree replacement work', which was to include my street. I'd been to a 'roadshow' by Amey and the council on Rustlings Road which was supposed to explain to residents what was happening with the trees. But they weren't asking us what we thought, they were telling us what they were going to do, what was going to happen, and expecting us to fall into line. I also met early campaigners from SORT [Save Our Roadside Trees], who were objecting to the planned fellings on Rustlings Road, which at the time was the main focus of public opposition. My feeling at meeting other opponents of the Council's policy was one of, 'If Rustlings Road people can say 'no' to it, so can we'. The difference for me being that on Wayland Road the felling was scheduled to take place very soon, and I had a feeling that actions as well as words were going to be needed.

Anne Barr: I had stood in the local Council elections a few weeks earlier in May 2015 and immediately after the election people began getting in touch about tree-felling in Sharrow Vale. Nicky told us about Wayland and Bowood roads which were next in line.

Nicky: The first time the felling crew arrived on Wayland Road with the chipper, they parked right outside my house. I just rushed out and I hadn't got any make-up on. I don't think I'd even done the flies of my jeans up. I looked like a wild woman; my hair is bad enough at the best of times. I rushed out and I stood under the tree and I said, 'You're not having this fucking tree!'. And then one of the young guys (an Acorn arborist) was laughing, and I said, 'What the fuck are you laughing at?'. Then I became a bit nicer, and I said, 'Look, there's a Council meeting soon about the Rustlings Road trees, can you wait until after that?'. By which point lots of other people had emerged from nowhere, including two guys who were fitting a kitchen floor at a neighbours' house and came and stood with us. I said to the arbs, 'I'm sorry I swore at you, but I don't want these beautiful trees to go'. On that first day people appeared out of nowhere. It wasn't a deliberate plan, but nothing got felled that day.

Calvin Payne: I'd attended my first meeting at the Town Hall on the issue a couple of weeks before. A few things struck me at that meeting; firstly how well attended it was, secondly how informed and knowledgeable the campaigners were, and finally how unsure and under pressure the councillors and staff appeared to be in the face of the opposition. I thought to myself, 'This is interesting,' but then towards the end of the meeting Nicky stood up and told everyone that the trees were scheduled for felling on Wayland Road very soon, and in no uncertain terms she told the Council that they would have a fight on their hands! That was the deciding factor in getting involved for me.

Nicky: Soon after there was another public meeting where councillors and officers put their case for the fellings and were confronted by a large number of determined and well-informed local residents. After both sides had presented their arguments, I told the councillors present that

I had seen the arbs off before and was prepared to do so again. The tree-felling had been rescheduled for two weeks later, with signs appearing on Wayland Road and the next street to us, Bowood Road, warning residents that both roads would be closed from 7am to 7pm from Monday to Friday to allow the work to take place. After the unsatisfactory outcome to the public meeting, I felt that an organised response was going to be necessary to prevent the felling of healthy trees on the road, so myself and several other campaigners spent a week knocking on doors and putting our plan of action to everybody who lived on Wayland and Bowood roads.

Calvin: We knew what we thought was right and what we needed to do, but we decided that we must know first what the opinion of the residents was on the roads. Anne [Barr] said that it needed to be a majority of them who were opposed to the fellings. I was very new to the issue and had no idea how people felt outside the meetings at the Town Hall.

Anne: Jenny [Hockey] from the SORT campaign came and helped us talk to residents on Wayland Road. It was the beginning of groups starting to form across different areas of the city. It was soon obvious that we had great support, with just one household supporting the Council policy.

Calvin: Just after nine o'clock on Monday 22nd June the felling crew arrived on Wayland Road to be greeted by a group of residents and supporters. Around a dozen supporters from across the city, including people who became very active in the campaign including Alison [Teal], helped local residents to maintain a presence on the street throughout the first day. We politely told the crew that we didn't trust the decision-making process and would be taking it further. They said they would report back to their bosses and someone would get back to us. And that was that.

Nicky: We had a table set up for hot drinks and we invited the arbs to join us in a pot of tea which they gladly accepted. It was all very friendly.

Calvin: The following day, the Sheffield City Council cabinet member responsible for the 'Streets Ahead' policy, Terry Fox, attended and addressed an impromptu street meeting where I asked him to, 'Respect the fact that local people were standing up for their local community' and to be fair, he said that he did. I think he thought he was being invited to meet privately with residents, but we made sure anything that he said was done in public.

David Kelly: Fox seemed to attend in genuine spirit. He did say that he respected what people were doing in standing up for their street, their community. Amey told the local press that we had endangered their workers with our actions but Nicky's tea party with the arbs and Fox's visit made that look silly.

Calvin: By Wednesday the crews had still not returned and we had media coverage for the third day running. Amey responded to questions by saying that the work would not happen that week and would be delayed. However, with the road closure notices still in place, I told them that we would stay on the street unless these signs were removed. Later that day two Amey workers walked down the road removing the signs, only asking that we didn't photograph them as they did so, which we were happy to go along with.

Nicky: I used to feel a bit out of the 'clique' on the road; most people had young families and I was a bit older, but suddenly that week I knew more of my neighbours and met so many like-minded people. Amey and the Council knew that whenever they tried again there would be someone waiting for them at all times.

***

Calvin: In response to the events of that week, and the growing opposition to fellings across the city, Sheffield City Council announced that 'bi-monthly tree forums' were to be held where experts from both sides met in public to discuss the issue. They only held two in the end, in July and September 2015. I was present on the second occasion, when the case against the felling programme was presented by Alan Robshaw of SORT. Alan gave the best presentation I have ever heard. He took down the Council's case for felling healthy trees line by line. I genuinely thought that it was all over after that. This turned out to be the last meeting ever of the 'tree forum' and it quickly became clear that the Council had no intention whatsoever of changing their policy.

Nicky: I was still a Labour Party member, so I went to a members meeting. I told them that we faced losing the local elections over the tree issue and said they (I was saying 'they' rather than 'we'!) needed to talk to the tree campaigners. I got a very frosty reception and was asked if I 'was even a member of the Labour Party' solely because of my opinion on the subject.

David: There was no need for any of what happened. Common sense, and a street-by-street policy could have solved any of the problems. When the Council ignored such compelling evidence presented to them, it made me wonder what else they had chosen to ignore or reject. I realised then that reasoned, mature analysis alone was not going to win the tree campaign.

Nicky: It was a celebration of ignorance and a complete refusal to listen to the facts.

Calvin: Some of the things we learned that week stayed with us: determined peaceful direct action, working alongside residents, attracting media attention, and being well organised. The basics remained in place. We started the week determined to stop the felling on the road and when we did, we realised that it was at least possible to do the same across the city.

June 2016

'I felt that I ought to repay a debt of gratitude to the protesters. This was something that I thought I could get involved with. An injustice that was worth fighting against.'

Ian Wilshaw, Resident

Bannerdale Road is a residential street in the Carterknowle area of Sheffield. The threatened cherry trees on this road had originally been paid for by local residents.

Ian Wilshaw: The first thing I saw was a sign on the tree outside my house saying that it was due for felling in the next two weeks. I didn't think that there was anything I could do about this as an individual. On the 6th June, I heard a commotion outside and went out to see a group of about half a dozen people under the tree. They asked me if I had received a Council tree survey and I said 'What survey?'. I remember the felling crews arriving at eight o'clock that morning and the police already being there.

Anne Barr: A few days before the felling crews arrived, some campaigners had delivered a survey from Save Nether Edge Trees [SNET] asking whether people had received, or replied to, SCC's Independent Tree Panel [ITP] survey.

The following evening, they went back to collect them from residents. 60 per cent of those received back said that they had never received any council survey. This would explain the very low and very surprising alleged lack of support from residents to save their beautiful cherry trees. I think that, as the survey was delivered in unmarked brown envelopes, people probably did receive it, but didn't open it and treated it as junk mail. During that day, more people brought their SNET survey responses to us in the street. On the first day there was a 6am call out. It was already light as I drove over from Nether Edge. My lasting memory of that morning is of two tree protesters frantically pedalling uphill to Brincliffe Edge in their pyjamas!

Jules Alexandrou: I arrived early and saw that they'd already felled one cherry tree. Chris [Rust] was already there under another. An arb said to him 'Can you please move out of the work site, so I can move my vehicle?'. Chris was very polite and said, 'No, I'm sorry about this, but you have had a good start today, after all'. We were confused about what we should do. At that stage nobody really knew what the law was.

Anne: When I arrived at Bannerdale Road I walked straight into the safety zone which was already set up around the second threatened tree on the road. I was told by Amey workers, 'You can't do that!'. 'I just have!' I replied. This was the first time I'd gone into a completed safety zone. At this point Dave Dillner, then chair of STAG, came marching down the road ranting about his court case [a bid for a judicial review into SCC's tree-felling programme which had been rejected, but was going to an appeal], and how my action was jeopardising it in some unspecified way. He told me to come out of the safety zone and I refused. After that he approached the Amey workers and told them that I wasn't part of the STAG campaign. I was so angry with him about this! Shortly after, I was joined by another protester, Calvin and a nice friendly neighbour who provided tea. More people arrived which meant that more trees could be defended as morning went on. Amey workers made frequent attempts to draw us away from the work zone, telling us we were endangering their, and our own, safety.

Calvin Payne: Dave [Dillner] had been instrumental in setting up STAG in 2015. I was at the formation meeting of the campaign which brought together a few different groups who were opposed to the Council plans. Dave brought a High Court case which had led to a pause in the felling initially, as an interim injunction had been awarded in lieu of the case being heard. That was a fantastic day for us, and if things had gone differently the case could have ended the whole thing there and then. However, it was the beginning of the law being used against us, and the start of the campaign needing to look to other methods. Trying to win our case legally was the right thing to do, but disrespecting good campaigners and tree protectors on the street wasn't.

Helen Kemp: Dave Dillner was shouting abuse at Anne Barr who was standing under a tree, 'Get out of there!' he was yelling, 'You're supposed to be delivering leaflets!' and then he was shouting at Anne, 'You're not even a member of STAG'.

Colette Cameron: One day during the felling on Bannerdale Road I spontaneously decided to do something to stop work. Further down the street I knew others were getting in the way and slowing things down. I stepped up and sat on top of a wood chipper, which immediately stopped work. Breaking social norms to stand up for something feels exciting and scary; I was full of both nervous energy and a sort of calm resolve to stay there for as long as I could. The police were called after I refused to leave, and we all waited for their arrival as the sun shone down fiercely. I had short conversations with Amey staff. I wanted to make connections, to keep things personal and friendly. I had noticed that some of the Amey employees were also finding this ongoing situation difficult. Rather than a 'them and us' standoff, I wanted to keep focused on the fact of our shared human experience. The police took their time to arrive. It seems they didn't want to get involved at the time. I sat on the machinery for three hours. A close friend stood by my side giving me vital emotional support, and the cheers of some local residents further encouraged me.

Anne: Eventually the police were called. This was the first time this had happened, as before Amey had always just packed up when confronted with protesters. The policeman adopted a low-level, friendly approach. He spent lots of time talking to Amey, us and his bosses. He didn't really know how to deal with it, but it was good-natured and diplomatic as the police were with us in those early days! We discussed Dave Dilner's court case between ourselves and decided that we didn't want our actions to impact upon it, despite the completely over-the-top abuse I had received from him earlier. We agreed with the policeman that when we were getting near to any potential arrest we would be given fair warning. The policeman was actually very supportive and spun out the time over a period of hours. Eventually, he came to tell us that it was time to come out as we could possibly be arrested. At this point we left the work zone as we thought any arrests might damage the campaign, but with hindsight I wish that we had stood our ground and not allowed ourselves to be talked out of it. We didn't have much legal insight in the early days and we were playing it by ear. I don't think we would have been arrested, but I didn't want to take that risk. Four hours later that tree came down, but they did not attempt to fell any others that day because there was strong support all along the road by this time.

Jules: The police came later. One of them threatened me with arrest but he didn't really seem to know which law I was breaking. Eventually, he told me that if I refused to move after being asked three times I would be arrested for obstruction, which I think he'd just kind of made up on the spot! At that time, I don't think that they [the police] knew what they were doing.

Calvin: It wasn't until a bit later in the campaign that we were more confident about the law in regard to what we were doing. It's one thing to decide to break the law on a point of principle, but quite another to not know that you are doing it. We knew we were morally right though, and that was our starting point for anything we did.

Anne: By then Andy Kershaw from BBC Radio Sheffield had arrived and was interviewing residents, some of whom were in tears, either from rage or distress. They told Andy that residents had paid for the trees thirty years ago and some of those who paid were actually there on the street. We walked up to give moral support to those who were guarding other trees.

This was when I first met Ian [Wilshaw] who was defending the tree outside his house. No further attempts were made to fell that day, but unfortunately they did get some more down as the week went on. The road is quite long which made communication between protesters difficult, and some of us couldn't return because of work. In those days only a few were prepared to do NVDA. Numbers greatly increased as the campaign went on.

Jules: Eventually after three days they'd taken the lot, all ten of the trees, and I remember thinking that if there are around three hundred trees in Nether Edge and they can do ten in three days, it's only going to take them a month or so. One of the things about Bannerdale Road, well, it was a bit deflating, and there were a few firsts for the campaign here – the police attending and threats of arrests – but it did galvanise people, and the residents were supporting us and coming out and giving us snacks and drinks, and I think we learned the value of engaging with the residents that week.

Colette: Although it was a stressful experience, I am pleased that I stood up for the natural world. I am still full of admiration for others who were able to put themselves on the front line time and time again. It was a short moment in the story of the protest, and just one of all the small steps of resistance.

Ian Wilshaw: I felt that I ought to repay a debt of gratitude to the protesters. This was something that I thought I could get involved with. An injustice that was worth fighting against.

2016

'*We just stood there hugging each other as the tree came down*'.

Sally Weston, Protester

Marden Road is a short residential street in the Nether Edge district of Sheffield. It is adjacent to an air-pollution blackspot.

Calvin Payne: For four months, one particular tree, a hundred-year-old London plane, was fought over as hard as any tree in the city at that time. The Amey felling crew had visited the tree at least five times and South Yorkshire Police had attended twice. On every occasion, the arbs left the tree standing when greeted by peaceful protesters. On the two occasions the police attended, they decided that the best course of action was to ask the felling crew to leave as the protesters were not breaking any law that SYP were aware of.

Helen McIlroy: We'd been guarding the tree for some time, starting shortly after the raid on Bannerdale Road in June. A lot of us were on a rota; Chris [Rust] had parked his car under it for a while too. On two occasions I was there before the crews turned up. Once because I'd followed them from the

Olive Grove Depot, the other because I'd been sleeping in my car keeping guard overnight.

Calvin: Helen had been camping out in her car under the tree in the summer. It was a real focus for people in the local campaign group. I'd first met Helen on Day One of our tree camp in Endcliffe Park the year before. I was preparing for the first night when someone I'd never met before walked across the field and said they were coming to stay with me. I was pleased that I was going to have some company for the evening, then she said she wanted to stay and camp for as long as I stayed there, which was fantastic!

Neil Meadows: The first time I was aware of the issue and the campaign was meeting Helen on Marden Road. She collared me one day as I was walking up the road and we started talking and she let me know what was going on.

Helen: A nearby resident also parked under the tree and one morning they called me and said that the police had told them to move their car, which was good for us as it gave us a few minutes warning! I think it was after that when I started sleeping there. I must have spent half a dozen nights in my car under the tree. I used to get there around 11pm with a sleeping bag in the car. When the crew found me there one morning they were surprised I'd got there so quickly but I don't think they realised I'd been sleeping there!

***

Helen: I slept there on the night of 1st November 2016. I woke up early and was sweeping up leaves to use for compost. The car was completely full of sacks of leaves. I thought I might as well do something useful whilst I was waiting there.

Calvin: I was sitting on some stone steps at the top of Marden Road opposite the tree. By this time Save Nether Edge Trees [SNET] had organised a rota of patrols around the area which began soon after dawn. Immediately after 9am an Amey felling crew arrived and began to set up traffic cones around the tree. The lead arborist, Carl Ellison, came straight out of the vehicle holding a piece of paper. He told me, 'Things are going to be different today. You're going to be arrested for this'. He handed me a sheet of paper containing the wording of a law I had never heard of, and which I was sure I wasn't breaking. In addition to organising a rota of look-outs we also had instigated a call-out phone to enable supporters to attend felling sites as quickly as possible. I started the call-out procedure and help began to arrive.

Neil: Whether it was a call-out or whether it was just seeing it out the window, I can't remember, but I do remember turning up on Marden Road thinking, 'This is a little odd', seeing all these people that looked like they knew what they were doing, and then the line of police vehicles.

Helen: I got the call-out soon after I had returned home with the sacks of leaves and by the time I'd got back, there was already quite a crowd gathering. Our group members were already standing inside the work zone; it was just a few traffic cones with tape across it in those days.

Sally-Anne Wickenden: When I got a call-out for somewhere close by I turned out to a tree protest for the first time. When I arrived there were already seven or eight people inside the work zone so I just stepped over the barriers too. It just seemed the right thing to do. I couldn't just stand by and let it happen.

Simon Crump: My partner at the time was a member of the SNET call-out group and when her phone went that morning she couldn't go to Marden Road straight away, so she asked me to go instead and said that she'd be along later. I'd heard a bit about the tree campaign through her, but I didn't really know what to expect. When I got to Marden Road, I saw a small group of people hanging around under a tree. I parked the car and joined them. Everybody seemed really relaxed and friendly and despite it being a freezing cold morning, it was fun. After about ten minutes a police car turned up and it started to get interesting.

Calvin: When the police arrived, led by Inspector Stubbs, he handed me a copy of the same piece of legislation that I had already been shown by the Amey staff. The previous police involvement in our protests had been cordial and mainly revolved around their perceived Health and Safety issues. This was the first occasion on which I had been threatened with arrest as well as being the introduction of Section 241 of the 1992 Trade Union and Labour Relations Act [TULCRA] which was to become such a major part of the story.

Sally Weston: When I arrived, there were already people under the tree inside the work zone. There was just something about the Amey crew that day. They were very smug, their attitude was one of 'problem solved'. My memory is that the police showed up in a series of vehicles including the dog van! There was quite a lot of discussion about what we might do as a group. I could see that Calvin was onto something and understood the implications. Inspector Stubbs looked very concerned and not at all happy to be there. He seemed really quite confused as to what was going on.

Calvin: The realisation that a piece of Trade Union legislation, designed by the Tory government to criminalise certain forms of protest, was going to be used in an attempt to settle a dispute between a Labour Council and its citizens was the deciding factor that today was the day to stand my ground. We'd had a few months of backing off by then, and I'd always been aware that this day would come. They decided it was going to be here and now by producing the Section 241 document. The police and Amey had made up my mind for me. We were going to either win our argument with Inspector Stubbs or be arrested. I didn't mind either way.

Sally: I was really cross that day! The then chair of STAG, Dave Dillner, arrived. He didn't speak to us, he talked to the police by himself and away from us. I was so cross with him! That's not how direct action ought to work. He tried to take control of a situation that he didn't have control of. He was patronising, treating everyone like school children. He didn't seem to be on our side.

Simon: This was all completely new to me, and the more I was told what to do by the police, the less I wanted to do it. I felt that the police were trying to intimidate the protesters and I didn't think we were getting enough support in attempting to prevent a beautiful tree from being felled, which up until that moment I had assumed was the whole fucking point of the campaign, in my wide experience of the last twenty minutes at any rate. It also seemed quite a coincidence that the police and the arbs were waving the same piece of paper at us. As I say, I was new to all of this and my view of events here was essentially an outsider perspective at this point.

Sally: Inspector Stubbs came up and asked 'Who's the leader?' and we said 'We all are!' and that empowered us!

Calvin: In total I was inside the work zone for four hours, and over three hours of that was spent in a stand-off with the police. Inspector Stubbs made several attempts to persuade us out from under the tree. He repeatedly stated that he 'didn't want to arrest us' and that it would 'serve no purpose' to do so. However, he could have withdrawn the threat at any time and found another way to resolve the issue. Personally, I had already determined to force the police into the choice; either acknowledge that we were not committing the offence or arrest us for it. I began asking if people were prepared to stand their ground and despite his eccentric appearance Simon obviously was. There were some in the campaign at the time who didn't understand what we did or why. To me it didn't take much thinking about. We were trying to stop a healthy tree being felled, we were protesting against the programme, and we didn't commit the crime we were accused of. If anyone still doesn't get that standing our ground was the right thing to do then I don't think they ever will.

Simon: I think by this time I'd decided that I was going to get arrested; it was actually getting quite tedious just standing there. I recall at one point asking Inspector Stubbs if it was warm in the police van. As I say, it was a very cold morning. At one point another protester David [Alcock] said to Stubbs, 'It appears we are at an impasse' and Stubbs replied, 'Yes, we are at a bit of a nonplus'. Eventually, after much pleading and giving us numerous 'final' and 'last' chances to move away from the tree, and Stubbs warning us that we 'wouldn't get any publicity out of this', we were politely arrested by a PC who put us into a cop car and took us to Shepcote Lane detention suite, which incidentally is just next door to *World of Sheds*, which is one of my favourite places in Sheffield.

Calvin: When we arrived at Shepcote Lane, to my great relief we were locked up in separate cells. I hadn't been arrested before, and it really is like on the telly. Shoelaces and belts are removed and the cell door is slammed behind you. There's a small, uncomfortable bench to sit or lie on, and a camera is always on you, even when using the loo. It wasn't personally traumatic or scary, however it could have been and would be to some. After a couple of hours, I was interviewed and had my fingerprints and DNA taken. I had nothing to hide, indeed I was happy to talk about what I had done and why. The only thing they kept asking me was whether I knew what I was going to do when I left home that morning, appearing to try and establish a planned course of action on my part. I couldn't have known of course, as I didn't know where or when Amey would arrive, or whether I would be there when they did.

Simon: Mostly it was just really boring. We were locked up in a tin box for eight hours, there was nothing to do, the bed was hard and the food was horrible. I had the cottage pie, which to be fair did taste a bit like a cottage. Room service was crap an' all. I was later told that we'd been locked up for so long because the offence they had dug up and were now trying to charge us with was so obscure that there wasn't even a code for it on the police computer. It was quite 'Kafkaesque': I was being imprisoned because they didn't know what to charge me with. Eventually they let us out at nine and they still hadn't charged us. I was under the impression that we would all have been arrested two by two and that the tree would have been saved, and I was very disappointed to learn afterwards that there had been no further arrests that day. Two days later the charge and a court date arrived by post.

Calvin: When I was locked up it did occur to me that it had taken four months, several attempts, a couple of police call-outs, weeks of patrols and lookouts, Helen sleeping in her car under the tree, a four-hour stand-off and two arrests for Amey to fell a single tree. Those occasions are always sad, but Nether Edge had over three hundred trees still scheduled for felling and it made me think how hard that was going to be for the other side.

David Alcock: It was a strange and cold morning, and most of us were trying to figure out the score. Despite losing the tree, Calvin and Simon volunteering to be arrested galvanised the whole war. It was the proverbial pivotal moment.

Sally: For me, direct action is quite exciting in a way, because you kind of make it up as you go along. You don't quite know what to do until you get there. If you're in a group, you kind of work it out and a group decision is the right thing to do. I think we kind of missed a trick here. Before Calvin and Simon were taken away, we discussed getting arrested in twos all day until it got dark. But when Calvin and Simon had gone, Dave Dillner completely took the whole argument away and suddenly it had all gone. I think the cops may have been thinking that too, hence the big van and all those other vehicles. In a way, I still don't know why we didn't do it. It all seemed to have happened too quickly, it all got cut off too quickly, which only happened because Dave intervened. They cleared us away and then they were up the tree lopping off branches. The police remained. Battling with my own thoughts, I really wanted to do more, but I hadn't because Dave sort of blocked it all off and made it a done deal. People drifted away, we were all shocked. I felt guilt and regret, not necessarily for not being arrested, but because we had been denied that discussion. We just stood there hugging each other as the tree came down.

17th November 2016

'All I could do was step over the tape that surrounded the tree. I stepped inside and then Freda stepped inside.'

Jenny Hockey, Resident

Rustlings Road is a residential street in the Ecclesall area of Sheffield adjoining Endcliffe Park.

Calvin Payne: In the early days of the campaign, Rustlings Road became a focal point for many people in Sheffield who opposed the Council's plans. It was the first road to be the subject of a petition which was presented at a monthly SCC meeting. The campaign group on the road [SORT] played an important part early on, by informing people about the Council's plans for the trees.

Jenny Hockey: I had a feeling that something was going to happen. After the September 'Tree Forum', the impression given was, 'We've listened to you, but we're going to do what we want with the trees anyway'.

Calvin: I got the call-out at five o'clock and when I left home in Nether

Edge at ten past five, there was a police car outside the house. When I arrived at Rustlings Road, the Ecclesall Road end was sealed off and the police were there. I was met by a PCSO who had previously been a regular 'neutral' visitor at the tree camp I'd set up in Endcliffe Park the year before (September-October 2015). She said, 'Calvin, it's Amanda from your "tree camp", can I have a quick word?'. She was trying to calm me down. 'It's five o clock in the fucking morning! Are you people still fucking neutral?' I shouted. As I walked along the road the first two trees had already had several large branches removed and work had clearly been going on for some time. I estimated that the arbs had been working for between one and two hours by the time I arrived.

Jenny: I was woken up by the police hammering on the door. I came downstairs and opened the front door to them in my dressing gown. My first thought was 'Someone's nicked the car' but in fact they [the police] were coming to nick the car. They wanted me to move my car so that the lime tree outside my house could be felled. I asked them what would happen if I refused to move the car and they said that they would take it away anyway. I kept saying, 'This can't be legal, this can't be legal!' All proprieties seemed to have gone out of the window. I refused to move the car. I went upstairs and put some clothes on (my jogging gear actually, because it was just lying around).

Helen McIlroy: Jenny phoned me before 5am and said, 'They're cutting down the tree outside my house'. I got up and out in minutes. There was a police car outside the house, which was interesting...

Jenny: When I went out the front it was an awful scene: arc lights, chainsaws and a huge trailer loading my car. What can you do? What could I do? All I could do was step over the tape that surrounded the tree. I stepped inside and Freda stepped inside. It was my only option after eighteen months of time-consuming campaigning. The democratic process had failed me. I knew that Freda was going away the next day for a long weekend in the Lake District and I said to her 'You don't have to do this, you know'. Then Freda's sister-in-law came in over the tape as well. She was going to a meeting that afternoon. 'I can't get arrested, I've got to go to Crewe', she said.

Calvin: It all happened so quickly compared to Marden Road earlier in the month. Three or four hours there, a few minutes here, and happening early in the morning in the dark. This was very different, only the law being used to ensure tree-felling was the same.

Jenny: The police gave us three warnings. On the final warning, a young man also stepped inside the cordoned-off area and said to the police, 'If you arrest them, you're gonna have to arrest me too' and as they did, Calvin handed me a card from Howells solicitors with Helen's details on it [Helen White, who had already become involved after the Marden Road arrests]. I could see men with chainsaws in every single tree as I was being led away. We were put into a police car and taken to Shepcote Lane detention suite. We were locked up from half past five until one in the afternoon when we were released without charge.

Freda Brayshaw: At 5:20am I was woken by loud hammering on my front door. Two policemen were standing there asking me to move my car or it would be towed away. I could barely take in what I saw down on Rustlings Road: blazing lights, a crowd of men in hi-vis jackets wearing head torches. I could hear the loud noise of chainsaws already in action all along the road. I realised that Amey, supported by SYP, had come like thieves in the night to fell our trees. When I'd flung on some clothes, I went down to the road. I saw that the tree nearest my house had been taped off. I stepped over a piece of tape to stand under the tree, unaware that there was a man with a chainsaw above my head. I stood there silently until I and the two others were arrested after a few minutes and taken to Shepcote Lane where we were held in separate cells for about eight hours. My memories of shock and anger I felt that morning after eighteen months of campaigning to save our city's beautiful heritage trees from felling remain vivid and disturbing.

Helen: By the time I'd parked the car and got to Jenny's house they'd already been arrested. The police had gone there to make arrests and they did it early to frighten us into submission. To sit in a cell knowing that where you live, your home, has been changed in that way, beyond repair, was awful I'm sure.

Julia Bodle: I got the call-out message at 5:30 and I had no idea what I was going to do. I didn't think that there was anything that I could actually do. I walked along the pavement and I could see that there was this cordoned-off area. And I just walked straight into it. I didn't really have any thought as to why. There were all these *No Entry* signs, but when I'm at work [as an obstetrician] there are always *No Entry* signs, but they don't apply to me, I just walked through with the arrogance that doctors have. A policewoman came up and told me that I couldn't do that, then she handed me a piece of paper and said that I would have to leave the cordoned-off area or I would be arrested. In my best Dipsy voice (and I do a very good Dipsy voice) I said, 'Surely that's not true?' So I stood inside the zone and started reading the piece of paper. 'Is it OK if I just stand here and read this?' I asked her. 'No it is not!' the policewoman said. 'If you don't move I'm going to arrest you'. 'That doesn't sound very nice' I said in my 'Dipsy' voice, 'That's not very pleasant is it?'. My daughter, aged eight, came out and said, 'Mummy, what are you doing?'. And I said, 'I'm standing under this tree to stop them cutting it down', and she was very pleased about that and came and stood under it with me. My husband came out to take her to school and they left me under the tree. 'Don't get arrested', my husband said as they left.

Calvin: After the events of the night and early morning I remember being interviewed a couple of times. The first time being on Toby Foster's breakfast show on Radio Sheffield just after 7am when I was asked about a police statement refusing to confirm the arrests, and I was able to say that three arrests had actually taken place, two of whom were retired women. After first light residents started to come onto the street from both Rustlings Road and surrounding streets, leading to disbelief and in some cases anger. The work then seemed to slow down with the worst of the damage already having been carried out under cover of darkness.

Helen: Bang on seven that morning they brought in the chippers, as per the rules on what time they could use loud machinery, and by then there were a lot of branches to chip. I spent a lot of time just milling around that day. We were shell shocked I guess. It was a very long day.

Jon Johnson: I turned up on Rustlings Road at 7am on the morning of the slaughter. I counted twenty plus-uniforms – as an ex-copper myself, I knew most of them – and was surprised to later hear the claim that only six officers were present.

Julia: At about 11:00 Andy Kershaw was on Radio Sheffield and we were standing on a garden wall. They were chain-sawing above our heads. I couldn't believe that South Yorkshire Police would let them do that. I jumped down into the safety zone to talk to the arbs. They wouldn't talk to me or tell me who was in charge. There was a Council observer there and I said to them, 'Somebody is going to get hurt!'. Nobody would talk to me, but they did stop work after I threatened to stand up in court and give evidence against them if anybody was injured. I was crying then, it was so horrible. It really wasn't about the trees for me, it really wasn't about the trees. It was about the people who claimed to be 'just doing their job' were so blinkered to the fact that somebody could get seriously hurt. At one point there were great big logs falling very close to us. I was genuinely concerned that somebody was going to die. It was around this time that Deepa [Shetty] appeared on the road and that the comment about them being 'Nazis' was made.

Calvin: I honestly think the whole problem with all of these people that we're talking about, is that they've all got a next person up who has told them, 'we can do what we want'. The police, the Council, the arbs, everyone's got a boss who just completely covers them. When you see [Amey account manager] Darren Butt on the street you know why the arbs were like they were.

Julia: At this point I still thought that everything was above board. Up until that day I had the utmost faith in the institutions of our country. I believed that the police were there to maintain the due process of the law, [in] a police force that didn't take sides and a Council that acted in the best interests of the residents of our city. That day turned my world upside down.

Calvin: By late morning there was a mood of defeat and shock that affected everyone, me included, and it took a while for us to regroup as the felling crews worked towards the last tree due for felling [referred to by residents as Ellen]. By afternoon the number of people protesting on the road was at its height and by the time there was only one tree left we had an advantage. Different tactics that were used to great effect by the campaign over the next couple of years had their origins that afternoon: trying to position a protester behind barriers, sitting on walls, occupying the gardens by the final tree (with permission). These were being tried and tested with some success at the end of a desperate day.

Zoe Borrowdale: When I went outside my house the arbs were already up the tree. I spoke to the two policemen, who told me that I would be arrested if I stood in the work zone, which included my garden. There was no real communication between tree [protesters] and residents so it was only after mine had been felled that I spoke to a neighbour, Luke, who talked about oversailing. We then rushed up to his house and stood in his garden. We had to ring my son's school to say he'd be late because the police said they would arrest him if he stepped off the garden into the work zone.

Alison Teal: I can't actually recall the time of day, but I found the daughter of the family who live closest to 'Ellen' [Rustlings Road trees were given names by campaigners] standing on the garden retaining wall, along with a few neighbours. Ellen had lost her branches on the side closest to the road, but some of her boughs remained overhanging the garden. An Amey arborist was there, trying to persuade the young woman that the tree was now in a dangerous state. He claimed he could not give any assurance that Ellen would not blow over in a strong gale and damage her family's home, thanks to his own butchery. A prominent campaigner was also present, and he appeared to be accepting of the argument the Amey worker was putting forward. He was attempting to persuade the daughter of the homeowner that she needed to go inside and allow the crew to fell Ellen. They had already taken seven apparently healthy lime trees and Ellen would be the last one they had hoped to take. I was as outraged as so many other people were, and I felt that it was important to protect Ellen. I intervened, and encouraged the daughter and others to remain standing on the wall. I spoke to the Amey worker and the prominent campaigner, and I said I thought we needed to have an assessment of Ellen by a person we could trust. There was some debate, but in the end the Amey crew gave up and walked away. A rota was established to guard Ellen night and day for a long time after, and I'm pleased I helped to ensure she's still there. Ellen was a small victory on an otherwise disastrous day.

Calvin: The scene and the size of the operation was off the scale of anything any of us had seen before. Because of Rustlings Road's symbolic status as a well-organised and publicised road, it appeared to many in the campaign that this was a heavy-handed attempt by a combination of SCC and SYP to break the resolve of the opposition and mark the beginning of the end of the campaign. Within days it was clear that the opposite was happening. The reaction to the news footage and reporting of events brought home the reality of what was happening to many, and the campaign's social media membership doubled within days.

Sheffield Hallam MP Nick Clegg [writing in the Sheffield Star, 24th November 2016]: In scenes you'd expect to see in Putin's Russia, rather than a Sheffield suburb, council contractors and police descended on Rustlings Road under the cover of darkness, dragged people out of bed to move their cars and detained peaceful protesters – all to chop down eight trees.

1st December 2016

'*And I was thinking, I'm on my own here. This is the day I get arrested*'.

Chris Rust, Protester

Raven Road is a residential street in the Nether Edge area of Sheffield. It is a short cul-de-sac lined with mature street trees.

Chris Rust: It was the day when Calvin and Simon appeared at Sheffield Magistrates Court to plead not guilty to breaking Section 241 of the Trade Union & Labour Relations Act [TULCRA] on Marden Road on 2nd November. There'd been a SNET meeting the week before, and between us we'd decided there was every chance Amey would make a raid while everybody was in town outside the court. So myself and Paul [Selby] decided to stay in Nether Edge and do our usual patrols around the area. When I was doing my circuit I must have missed Raven Road, because when I drove past it again Amey crews were already there setting up. 'Oh shit', I thought. There were two things on my mind then, one, to call for some back-up, and two, to go into the work zone and assess the situation. Phone reception in that part of Nether Edge is atrocious. I tried to ring the call-out phone and then realised that I already had the call-out phone. I finally managed to get through to someone who was amongst the hundred or so campaigners who had gathered outside the court to support Calvin and Simon.

Calvin Payne: After a short hearing to make our 'Not Guilty' pleas and to set a trial date, we came out of court and there were dozens of reporters and around a hundred supporters. As we were speaking to the media, we heard that there was an attempted felling in Nether Edge just a few streets away from the scene of our arrest. I never thought they would be so obvious and stupid as to attempt a felling in the area at the same time as our appearance in court.

Chris: There were two threatened trees on Raven Road, a knackered old thing which nobody objected to being felled and a lovely tree which was leaning slightly into the road. If it had been a busy road this might have presented a problem, but Raven Road is a short cul-de-sac. Amey had begun to set up around this tree. There were three or four residents out on the street already who were objecting to the tree being felled, and we knew at that point Amey would be calling the police. 'Hi guys, I'm sorry to spoil your day', I said (my usual line) as I stepped into the work zone. And I was thinking, 'I'm on my own here, this is the day I get arrested'. The Amey guys did call the police and I believed that the next thing that would happen is that the police would come and I would be threatened with arrest. The only thing was that I was absolutely desperate to go to the loo. One of the residents replaced me for a few minutes and then I went back inside the zone again. Then the police came and I was told it was the same two policemen who'd been assigned to the rally outside the court!

Calvin: One of the police officers from the court rally arrived on Raven Road about five minutes ahead of me!

Chris [speaking to the media at the time, taken from a YouTube film]: If necessary, yes, of course I'm prepared to be arrested. We have to challenge the cases that are being brought under the Trade Union law. Personally I feel that those arrested already have a strong case, a strong defence. For me today, the challenge is that if they continue to threaten people with this law without charging them, then it becomes frivolous action by the police, an abuse of their powers. So it's quite important that when threatened with arrest we don't just say, 'OK, I'll go'. If we knew getting arrested would lead to a conviction we might then say, 'What's the point?', but in this case we've got to test the law, challenge whether it's being used rightly, which it appears not to be.

Calvin: We were able to tell the media who were at court that they had the chance to witness a live piece of direct action. Sheffield City Council had inadvertently organised the perfect media event for us. As the campaigners left to head to Raven Road we were followed by reporters, making a convoy of vehicles heading out of the city through Nether Edge. There were several cars and a couple of bicycles leading the media to the site.

Chris: Suddenly the street was filled with campaigners, supporters and all the media who'd been outside the court. It was perfect, we couldn't have organised a media event like this without Amey's help! There was a stand-off and I stood under the tree in my baby blue puffer jacket (one size too big), enjoying my moment of glory.

Calvin: It was a really good day in the end. Eventually the Amey supervisor turned up and told the crew to pack up. Thanks to Chris that tree is still there. To have the awareness and organisation to ensure a good crowd inside and outside the court and then at the same time prevent a felling in the area was great! Although we didn't know anyone on Raven Road itself, two of the neighbouring roads, Briar Road and Ladysmith Avenue, were being looked after by residents Jenny [Saul] and Chris, who had organised the campaign activity on their roads. It was a strong area for Save Nether Edge Trees.

***

Jenny Saul: Briar Road has a great spirit, a residents' Facebook page, street parties, the lot. That definitely helped us get organised over the tree-felling threat. In 2016, I saw an Amey guy measuring things on the street.

Someone from the road contacted the Council and was told which trees were scheduled for removal. We were then told that it wasn't definite and we shouldn't have been given the figures! I delivered information about the Tree Survey that the Council was about to send out, preparing residents to look out for it. As a result the road returned a 100% opposition to the plans on a 40% turnout, considerably more than most other streets. Somebody else on the road started the tree group on Briar Road. I think it was in January 2017 that we began organising ourselves. I remember some people were worried about arrests, being teachers or similar, but others were worried because they had been arrested several times before and had 'previous'! It's an awesome street! Some hotter heads had to be calmed down a bit as there was talk of buckets of urine being kept next to upstairs windows. We even had a conflict-resolution specialist amongst the group who came in very useful at times!

Calvin: Sometimes the tree issue developed a sense of community on a street through joint endeavour, sometimes it was there already and greatly helped the cause.

Jenny: The one big attempt on the road was when they set up around the tree outside my house and a few of us occupied the front garden close enough to the tree to prevent them from working. Alison [Teal] had come on our Facebook page and said that we might not always be able to save the trees, but promised that come what may, she'd be there to help us try and prevent it, a promise that she kept to the letter.

Calvin: It's been a pleasure to have a local Councillor who not only speaks up for you but literally stands with you fighting power and wrongdoing. On a couple of occasions, I've been able to vote for someone who has stood in the dock with me. That wasn't a hard decision to make!

Jenny: The other encounter with Amey outside my house was post-injunction, when they came looking to remove bat boxes. I went outside and the arb said, 'You can stand there, but I'm going to set up the work zone around you, read you the injunction, and photograph you standing there'. I tried arguing with him that this wasn't a tree to be felled as the injunction says, but I wasn't all that confident about it. Others soon came and stood under the other trees on the road, but I was alone in a completed work zone. I wasn't really ready to breach the injunction without a disguise. Russell [Johnson] was there and he advised that I might be in breach. So I left the work zone and joined the others in getting as close as possible to the trees.

Calvin: There were plenty of grey areas around the terms of the injunction and anything that wasn't tested in court remains undecided. Certainly the arbs and security weren't usually the opinion to listen to regarding the law! Sometimes discretion was the better part of valour, especially when the tree could be defended without putting yourself in that situation. Briar Road, Raven Road and Ladysmith Avenue were roads that were well organised and with large majority support for the campaign and that meant Amey didn't try too often. There are some roads which may seem relatively low-key, but that's because they were ready and the other side knew it. They were victories as much as anywhere else.

February 2017

'I honestly felt bereft, almost like I'd lost a friend, and I suppose I had.'

Elizabeth Gash-Wales, Resident

Chippinghouse Road is a tree-lined street in Nether Edge adjacent to an air-pollution blackspot.

Paul Brooke: On the morning of Monday 6th February I was on my way to work. I'd seen something on the STAG Facebook page saying that they'd come for a tree on Chippinghouse Road, so I thought, 'OK, I can pass that on the way, pull over and show my face, lend a bit of support for half an hour, then I'll go to work'. So I pitched up to do just that. Simon [Crump] was already under that tree on Chippinghouse Road, Inspector Stubbs had just arrived, and he was telling people that they would get arrested if they didn't move out of the barriers. They messed about for around twenty minutes and I just stood outside watching. There was the usual discussion about, 'You have to be outside the barriers, and if you don't you'll be arrested under 'TULCRA Section 241', and gradually everybody moved outside of the barriers.

Simon Crump: On the afternoon of Sunday 5th February, Alison [Teal], Calvin [Payne], Chris [Rust] and I met at Chris's house to talk about what we might do in terms of tactics for the coming week. We knew that the felling crews were coming on the Monday because the signs had gone up already all along Chippinghouse Road. After a bit of discussion, we decided that the best way to delay the felling, at least in the short term, was to wait until the arbs had set up and were actually in the tree, then we would step over the barriers into the work zone. We knew that they would then call the police. After that, the plan was that most people would leave the work zone so that they would be out of earshot when those who remained in the zone were given a final warning by the police to move or be arrested. Then those people would step out of the zone, the police would leave and the arbs would set up in the tree again. As soon as they'd done so, the people who hadn't heard the warning would step into the work zone, the arbs would stop, the police would be called and the whole thing would start all over again. It wasn't a great plan, but it was all we could come up with, and we reckoned we could keep them busy all day like this until it got dark. When I got home that night, I was very surprised to see a directive on the STAG Facebook page, issued by the then STAG Steering Group, telling us not to get arrested in the coming week because 'they' were hoping to enter into discussions with the Council. My personal feeling was that it should work both ways and that there should also be a pause in the fellings while these discussions took place, but it was clear that this was not going to happen. I was lucky to get the first response on the page to this new directive. It read simply; 'FUCK YOU STAG'. And a lot of people 'liked' that comment! Anyway, on the Monday morning we put our genius plan into action and it worked really well for at least twenty minutes, but when the police came back for a second time Inspector Stubbs had obviously worked out what we were doing. And he was very cross.

Paul: So people left, but some of us had avoided hearing any warnings, so I thought, 'OK, this sounds like a bit of fun', so I went back inside the work zone and Stubbs had to come back because people had gone back inside the barriers. So then we were told that we needed to move outside the work zone and by that time there must have been probably twenty of us. We moved outside and then we lined up, all touching the barriers as close as we could get. The young arb then started to go up the tree. He got about halfway up, then he looked around and said, 'This still isn't safe. I can't work with these people this close', (those were the days!). So the arbs on the ground then wanted to start moving the barriers. Stubbs and two PCSOs were there at the time and there was a lot of jeering and stuff because the police were helping them to move the barriers. So the police started to move the barriers back and the aim was to block the whole road by putting barriers right across it. Then Stubbs said, 'The safety zone is inside these barriers up to the pavement, so you can stand on the pavement on the far side of the road, away from the tree'. There were lots of people commenting on the fact the police were acting as Amey's own security force. Stubbs said, 'I'm authorised to be able to close the road if I want to, so I've closed the road, that's where we are. You can stand on the path'. So gradually everybody milled around, shuffled, took it slowly and eventually got onto the pavement on the far side, and the young arb started going up the tree again. While I was standing on the pavement I'd spoken to a woman who mentioned that as a child she'd collected conkers off that tree, and that her children had collected conkers off that tree too. I stood there on a cold sunny day looking at this big tree, already having had a look at the pavement and the damage the tree was supposed to have caused, and thinking, 'This is ridiculous!'. The kerbstone was no more than an inch out of line. I just felt so powerless, standing there thinking, 'How am I going to just stand here and watch them cut this tree down?'. I thought, 'Right, I've either got to leave or do something'. The arb was starting to go back up the tree and just as he got there and called for his chainsaw, I don't know why, but I just went, something inside me just went, 'I'm not having this'. I didn't speak at all. I just walked across the road, sat down cross-legged on the pavement and leant my back against the tree. I didn't say anything to anybody. I sat there and deliberately just crossed my arms and looked at the ground. I thought, 'This is my own private thing, and if I get arrested, I get arrested. That's it, I'm just going to sit there.' As I sat down I thought, 'Well I'll probably have to stick with this now, yeah, OK, I don't know what I've done, but I've done it now. I'm happy with what I'm doing'. I just sat there. I heard someone else, it might have been Simon [Dewick], shout, 'Yay!', and somebody called something else out. I heard

Alison [Teal] at one point shout, 'We can't leave him there on his own!', and then Alison joined me, Simon [Dewick] joined me, I think they were the first two that came over, and then gradually the others came over, Jeremy [Peace] came over, and suddenly as I put my head up, there was a bit of cheering. I looked around and there were at least six, then seven, then eight of us.

Martin Pickles: The day of the mass arrest I think we'd all got to a point when we'd had enough. I just thought, 'I'm going to go and sit under the tree', but Paul [Brooke] beat me to it! I was sitting there thinking it would be a good idea to phone my wife, who pleaded with me that it might not be a good idea to get arrested as I was picking up our kids from school! I apologised profusely to Paul and the others as it looked like I was copping out, but the kids had to come first.

Paul: I knew about Calvin and Simon's arrest on Marden Road, but I didn't know all the details. I was talking to somebody on the day and they said, 'Oh, that's the same charge that they used before', and the other cases hadn't been dropped by then. The other thing that was on my mind was that if I was going to get into trouble over a tree, if I was going to get done, it'd better be either the one outside my house or a tree that I felt was a glorious tree without any justification for felling. The tree I was under was outside the house of the chair of the Independent Tree Panel, so I thought, 'Well, actually there may be some press mileage in this, so it might be worth taking the risk of getting arrested'. I remember sending my wife a text message just saying, *'I think I might be going to get arrested today, sorry!'* Yeah, so for me it was just an emotional, 'Let's see what happens' kind of feeling.

Jeremy Peace: I was on my way back from the shops; it was the first protest I'd attended and I'd left the slow cooker on at home. When I saw people going to sit under the chestnut tree, I just sort of got carried away with the moment. I thought I'd best get stuck in quickly and then it began to dawn on me that I wouldn't be going home any time soon!

Kate Billington: I thought I'd go down for an hour to lend my support. I left the heating on and went to see what was happening. I didn't believe that the horse chestnut tree was dead, diseased, or dying. I didn't have any proof other than what my eyes and my gut told me. I also didn't understand how trees could be felled so easily in a Conservation Area. I still don't know quite how many laws those chancing bastards got away with breaking. We were all outside the barriers and at what felt like an important impasse a guy in a hat (we were all in hats – it was bloody freezing!) who was standing next to me just said, 'Fuck it', and broke through and sat down, arms folded, under the tree. I glanced around and saw other people making to move and thought, 'Yes, fuck it!', and joined him. By the time the dust had settled there were seven of us. I only recognised one other face, Alison Teal. Her presence under that horse chestnut reassured me that what I was doing was righteous and not daft. Inspector Stubbs droned on for ages, it genuinely seemed like he didn't want to make this mass arrest. It was all very calm, and we just sat there.

Simon Dewick: On Monday morning, I walked down Albany Road as usual, and turned onto Chippinghouse Road to witness the gathering of police and Amey workers in yellow vests and hard hats. Behind a set of barriers there was a group of people asking the arbs why they intended to cut down a perfectly healthy tree. I realised it was a tree that I used to collect conkers from for my children over the years. I remember feeling angry and that it was somehow acceptable to cut down a perfectly healthy tree for no other reason than it was displacing the kerb by a small amount. I was impressed with the patience of the gathered crowd who were engaging a senior police officer with intelligent and reasoned arguments why the felling should not proceed. I agreed that, apart from its aesthetic beauty, this tree provides oxygen, takes away carbon dioxide, its canopy protects us from the sun and soaks up excess water. I asserted that I thought I lived in a Conservation Area and that it was not right to kill the tree. It was clear that whilst the policeman listened to the arguments he had no intention of averting the planned destruction, citing the need to clear the area so workers could get at it.

Paul: Well, there was then quite a lot of sitting around and waiting, and then chatting with Stubbs. He initially brought out the sheet of paper with the 'TULCRA (Section 241)' legislation written on it. He asked us if we wanted a copy of it, so I took one, and then he said, 'I'm warning you that this is the law that you're breaking, and if you don't move from this work zone, you will be arrested. This is your first warning'. I think he told us that he would give us five minutes to think about it. So, I was sitting there reading through the sheet of paper. I've got thirty years experience in adult social care. I worked as a senior manager for Sheffield City Council up until a couple of years before then, and I'm used to trying to make sense of complicated stuff on bits of paper fairly quickly. I read through this 'Section 241', and I could see the way it was laid out on the sheet, it was in two halves. It says, 'You're in breach of this Act if you do this, this or this and you do it like this, like this, or like this'. I thought, 'Ah, that's funny, I'm doing that bit, but where it says 'and', not 'or', it says 'and', so I've got to do both parts of this to break the law. I said to Inspector Stubbs, 'I understand that bit, but what am I doing wrongfully?'. He says 'It doesn't matter if you're peaceful, it doesn't matter if you're quiet, by doing that you are breaking that law', and I said, 'No, no, but it says "and", in fact it was in bold. I said, 'Can you tell me the "and" bit that I'm doing?'. He said, 'It doesn't matter, it doesn't matter', and I thought, 'I think it probably does matter!'. I think at that point I thought, 'Nah, I'm seeing this through, because actually I don't think I am breaking the law, so I'm going to stick with it. I'm going to go through with it'. We were all chatting between ourselves, and I think by that moment maybe there was just a collective attitude of, 'We're just going to stick with this and see where it goes'. What was funny is that I only became aware of STAG's 'no arrests' directive two or three days later. Simon [Crump] said that he misread the memo's 'none to be arrested' as 'nine to be arrested'. I remember that. What was interesting was that I didn't know anyone else who was sitting under the tree with me. So it was kind of odd, because it was totally unplanned, it wasn't coordinated, none of us had spoken in advance. None of us said, 'Look, should we get done for this one?'. Nothing about it seemed right and fair, and I felt a sense of, 'Actually I'm not doing anything that I shouldn't be doing'. In fact I was positively doing something as a concerned citizen of the city that I *should* be doing. What I was doing was a positive act rather than a protest for protest's sake. I actually felt I was doing something positive. I hadn't been involved in it before, but the process of getting arrested was that we had to wait for the police vans to arrive, with different police officers who hadn't been involved in removing us from the protest. Two officers came over, and they were charged with actually carrying out our arrests. When we walked into the police station at Shepcote Lane, we were all chatting and it was very relaxed. They didn't have enough holding cells because they put you in a little holding cubicle at first (those tiny little phone box things) and there's a photograph of Maggie [Mark] sitting on my knee and the two of us laughing and joking. Maggie was just in hysterics and found the whole thing absurdly funny. She could not stop giggling and laughing! A speck of blood on the wall just cracked us up, 'Oh! maybe this could be serious'. But she was just so funny. Then you are locked in a cell and you lose all idea of just how long has gone by, is it two hours, is it three hours, or more? You kind of lose track. It was all pretty uneventful in terms of the custody suite bit and in the interview Helen [White] from Howells Solicitors represented us, gave us some advice on what we should and shouldn't talk about, etc. We were there for about eight hours. We dropped by Chippinghouse Road on the way back, taking Jeremy and Barbara back home. We were taking them back so I said, 'Well let's just pull round and see if there's any chance the tree is still there'. Of course the tree was by then just a stump, and that was really, really disheartening. I did then question whether my actions had made the slightest bit of difference, who knows? What's interesting looking back and reflecting is that the 'Battle of Chippinghouse Road' made a huge difference. An absolutely huge difference.

Kate: I introduced myself to Jeremy [Peace] in the police van, not knowing that that was the start of a good friendship. The police station duty officer who took my belongings told me quietly that 'between you and me' he was on our side. Unlike the automaton who took my biometrics and the worst ever photographs. My one phone call was to a neighbour to turn off my heating and feed the cat! I got to sit in a police interview with a good cop/bad cop kind of setup and was instructed to say, 'No comment' to everything, which felt really similar to how it looks when you see it on the telly. I have to say, all this was manageable because of the safety in numbers aspect of it. I had six other people to turn to in the future, however it panned out, and the fact that Helen from Howells was already at the custody suite when we arrived – not a woman to mess with! Paul's partner came and took us home from Shepcote Lane, eleven hours after I'd 'gone out for an hour' that morning!

Simon Dewick: I watched a man I didn't know walk slowly towards the tree and sit under it with a smile. My adrenalin started to rise and I said to myself, 'Yes, enough is enough', and I joined him and five others. I didn't know anyone else there at the time but I felt a tremendous sense of pride that I was with others who were prepared to try to protect the tree by peacefully sitting under it. The police responded by stating their intention to arrest us all if we didn't move. I thought that I would be able to remain around the tree to prevent work, but was told that a van would soon arrive to take us to the station to charge us all. Seven of us were separated from the others, and all seven of us stood our ground to assert our right to peacefully oppose this act of vandalism. All seven of us were then arrested. It was stressful to be held in custody for over eight hours, but I did not feel guilty of any crime. Our arrest did not save the tree, but continued to strengthen the roots of a campaign that has succeeded in protecting healthy Sheffield trees. I feel proud to have taken part as one of the 'Chippinghouse Seven'.

Kate: My dad sent me the money for my legal fees and also a card he made which says "Resistance is never futile". He's 83 and makes me proud to be a Billington – though I absolutely must point out that the erstwhile Sheffield Council's Director of Culture and Environment, Paul Billington, is absolutely NOT related to me nor to my father!

Jon Johnson: As I drive around Nether Edge, my heart sings at the sight of so many magnificent trees. Each time I am reminded of a different abhorrent episode in the three-year fight to keep those trees there. A fight that has been undertaken by ordinary people; normal, everyday people, most of whom previously wouldn't have kept a library book out too long. As a former police officer, I have a strong moral compass. Those memories are also of a Council whose own compass hasn't just wavered, but rather come off its gimbals completely and rolled away into a very dark corner.

Calvin: Jon used to come up behind me and place a strong hand on my shoulder. I'd say, 'You've done that before Jon!'.

Jon: I've seen former colleagues bully and harass citizens of this fair city, sorry, *formerly* fair city, on the direct orders of a corporation, aided and abetted by a bunch of abject conspirators who took time out from mismanaging the rest of the city to focus their hatred on the 'middle-class' residents of Sheffield's 'leafy suburbs'.

***

Simon Crump: On the morning of Wednesday 8th February, Alison, Natalie Bennett [former Green Party leader], and I were standing on a driveway in Crescent Road, which is actually looking out on Chippinghouse Road because the house is on the corner. I'd been there the day before with Graham [Turnbull] and between us we'd successfully seen the crew off, because we were on private land with the owner's permission, and we were close enough to the perfectly healthy lime tree to prevent it from being safely felled. The police had tried it on with us, but had eventually given up and the felling crew had left. Next day, the police were in attendance again and Inspector Stubbs wasn't having any of that. He asked to see the written permission we had to stand on the driveway and we suddenly realised that none of us actually had it with us! I went round the front to see if the resident was at home and luckily she was just coming out of the front door on her way to work. We found a piece of paper and biro and she hastily scribbled us a new garden permission while resting on her black wheelie bin. I went back into the driveway and handed the piece of paper to Stubbs. He didn't seem that impressed by it and warned us that if we didn't move we would be arrested. All three of us were very keen to legally test garden permissions and we were pretty sure that we couldn't be arrested for standing in someone's driveway with their permission. There was more debate between us and Inspector Stubbs about this, but by this time he just seemed to be making stuff up. We asked him which law we were supposed to be breaking and he replied, 'The law'. Eventually, he put us on a five-minute warning, move or be nicked. It was difficult to know what to do. Alison was on bail, I was on bail and Natalie had a conference to speak at the next day. Eventually we decided to see if anybody was prepared to take a risk and replace us. There was quite a crowd gathered by this time and so I asked around if anybody was up for it. I felt like a magician's assistant selecting two lucky members of the audience to come up on stage and be sawn in half. I was spoilt for choice as it happened, and within a couple of minutes Kezia [Richardson] and Gemma [Lock] stepped in. After a brief telephone conversation between Stubbs and the Council they were duly arrested. Before they were led away, Stubbs promised Kezia and Gemma that they would not be locked up for the full eight hours, but that turned out to be untrue.

Gemma Lock: I arrived at the Chippinghouse Road protest on 8th February at about five past ten. I was asked by another protester if I was willing to get arrested, I said 'Yes' and went to stand inside the garden of a campaign supporter. She gave Kezia a written note giving us permission to stand in her garden. I was not given any other warning prior to being told just once that I was going to be arrested if I did not move. I was then arrested at about quarter past ten along with Kezia. We were taken separately to a police van. We were separated and taken to Shepcote Lane custody suite. We each had an officer escort us to the van and then while we were 'processed' at Shepcote Lane. My possessions were taken away and I was put in a cell after having my fingerprints, DNA and photo taken at around quarter past eleven. I was then taken to see a nurse to prescribe my Ventolin inhaler for my asthma. I hadn't been offered any food, so I asked for lunch at around 2pm. I had a visit from two mental-health workers around half past two. I asked them to organise some food. I received a bowl of chilli at around 3pm. I was in the cell until about quarter past five. I was then taken to see my solicitor Helen White and prepared a statement. Then I was interviewed by the police and was bailed with the condition of not going inside a work zone when tree protests were occurring. I was given a restricted-bail sheet stating this. The police initially tried to set not protesting as a bail condition, but my solicitor refused that on the grounds that I should be allowed to peacefully protest. The police then took Kezia and myself back, in my case to my van near the site of our arrest, at around 7pm. On the way home the police were very friendly, and asked us lots of questions such as

'Who is your leader?' and 'How do you communicate with each other?'.

Kezia Richardson: I'd become aware of the tree issues recently, and after doing the school run I wandered down to Chippinghouse Road, where several mature trees were earmarked for removal. Several people had been arrested earlier in the week, and it was clear that SYP and Amey were in a bit of a flap about it all. That morning, Amey was trying to kill the mature horse chestnut near the corner of Crescent Road. That tree was arguably the finest conker tree in the area, and my children loved it. The tree was overhanging a garden on the corner and several protectors gathered in the garden, while Inspector Stubbs tried to exert his authority by waving printed copies of a Thatcher-era anti-Trade Union law about. The owner of the house was swiftly contacted and brought down written permission for protectors – and not SYP or Amey – to stand in the garden under the overhanging branches. This got right up Stubbs' nose, and after a panicked call to the SYP solicitor, he declared that we were risking arrest by preventing a worker from going about his business. We were rightly incredulous that SYP would stretch legislation so far to protect the interests of the Amey/SCC contract. We decided to challenge this. As I'd handed over my kids to their dad for the next few days, I had a bit of time to spare. I accepted Simon's request to remain under the tree to see out the process, with another protector, Gemma. By this time a crowd and the press had gathered around the barriers, and as Gemma and I were read our rights and led to the police van, the booing and jeering was heartening. I'm not a public speaker by nature, but I did address the crowd briefly and I think I got on Look North. After being kept in a cell for about eight hours, and interviewed at length, my partner in crime and I were given a lift home by the same officers. They were desperate to find out how we organised the actions. I did manage to nick a roll of police barrier tape out of the van as a souvenir, and no-one else was arrested under that legislation. The garden permissions became quite a useful tool for protectors too, so even though we lost that particular tree, I feel like it was a good day's work, all told.

Paul: On Thursday 9th February there was a message on Facebook that the crews were out early doors on Chippinghouse Road. I got there around a quarter to eight and the crew was there with a few barriers around the tree outside a house near the junction with Abbeydale Road. They were there to fell that tree. If we didn't do something, it was going to be gone. Having been arrested for allegedly breaking it three days previously, I'd become very interested in 'TULCRA Section 241' and seeing as I was a contractor [a joiner] myself, I was wondering how it would apply if I was the one being prevented from working. At this point Stubbs appeared on the scene wearing his 'civvies'. He was on his way to the cop shop and said he had just called in to see what was happening. 'Are we going to have the same stuff today?', he asked. 'Hypothetically, what if a worker is already working where the felling crew want to work?', I asked him. 'I'm asking that question because I'm supposed to be painting the resident's gates today'. 'You've got a point there,' Stubbs said. 'Also, if they put those barriers up around me while I'm working, that will criminalise me and I will be in breach of my bail conditions not to enter or remain in a safety zone set up around a tree' I said. Stubbs disappeared for a minute and when he came back he said, 'I've told them to stop setting up'. Out of earshot of Stubbs, I said to Chris [Rust], 'I think we've got something here. I think I need to speak to the owner'. I knocked on her front door and she opened it. 'I've told Stubbs I'm painting your gates,' I said. 'Well yes absolutely, they need painting! I'm happy to commission you to do that'. I thought I'd better think about getting something in writing, meanwhile I brought the van round, parked it under the tree and started setting up. I played up the comedy a bit. I took one of the gates off its hinges, put it on trestles and got my radio and flask out of the van, and started complaining about the protesters getting in my way! Ironically, I did have a real job to go to, and so after half an hour or so I left. The reason I gave was that I'd gone to buy some paint. I went to my real job and kept an eye on Facebook updates on my phone. By this time Sian [Thomas] was livestreaming from inside the house accompanied by Alison [Teal]. Stubbs and another copper were in the sitting room putting pressure on her to provide a date when she had commissioned me to do the work. Then an unknown number came up on the screen and I answered it. It was Inspector Stubbs! 'Hello Inspector Stubbs, how very nice of you to call! Yes, I'll be back in about 45 minutes'. Stubbs' legal advice was that Amey had instigated the work before me and that I must give way to them. He wanted to know when the homeowner had commissioned the job, and I was vague to the point of deliberately misleading the police. 'Oh I don't know, it was ages ago, you know how these things are'. Stubbs thanked me and rang off. A few minutes later he called me again. 'How long will it take you to finish the job?'. 'I'm going to need all day tomorrow'. 'So you won't be on site on Monday?'. 'No I won't, I'm quite happy to agree to that'. He thanked me again and rang off. We both knew that it was bullshit. And that tree still stands today.

***

Elizabeth Gash-Wales: I received a letter at the beginning of December 2016, about the felling of Lonesome George, 'my tree'. I was hoping the debacle would be long over by now and that SCC /Amey would have come to their senses and it would never happen. How could they take such a beautiful, healthy tree? I had read the Independent Tree Panel report, which deemed it to be healthy, and as most people know, 'George' produced hundreds, possibly thousands, of conkers every year. I moved into my house, directly opposite Lonesome George, in September 1988, with my then two children. They have grown up with that tree. It's been a source of fascination and fun for years. My son rescued baby birds that had fallen from the tree, they watched squirrels running up the trunks. In fact he climbed that tree many times, on one memorable occasion attaching a line from the attic window to the tree, which he and his friends sent their 'Action Men' across on. When my youngest daughter was born, Lonesome George was one of the first things she saw. Lonesome George was a source of many things for my children, nature talks, activities, and we always ended up with bags full of lush, shiny conkers, which they'd take to school to share out. Living in such a busy, noisy, built-up area, my tree gave me greenery and nature. Literally through my front window I could see the seasons change, it gave me a sense of peace and calm. Going back to that dreaded letter, I decided to use two weeks of my annual leave to try and save him and the other trees further up the road. I was out daily, like many others, standing in the cold. As we know, that week on Chippinghouse Road was emotional and utterly exhausting. The protesters fought with all their might to save the trees. My daughter, Scarlett, who lives overlooking the Olive Grove Road depot, would message me when the trucks were leaving and I would pass the information on. I still remained hopeful, but the day came, far too quickly. I set my alarm for 5:30am, but I didn't really sleep. I just kept looking out the window. The evening before, Amey workers were putting notices on cars, warning they'd be removed if they were left on the street. Myself and another neighbour took my dogs out for what might be their last chance to pee up Lonesome George. We were called 'stupid cows' by one of the Council workers, as the notices were being removed from cars as fast as they were put on, and he presumed we were doing it. I had a dog on each hand, how could I possibly have done that?! The same Amey guys arrived very early and parked shining their headlights into my front room, which is when I put the alert out and hoped someone was awake and would come as I felt very alone at this point. Initially I felt like nobody was coming, it seemed to take ages. but then they started to arrive. It was a bit like that film, *Field of Dreams*: 'If you build it they will come'.

Simon Crump: There's a photograph of us all under Lonesome George, taken early in the morning before the felling crew arrived. It's a lovely, hopeful picture of us all together and we look like an overgrown gang of kids ready for anything. Looking at it now, it makes me sad.

Elizabeth: Steadily more and more people had arrived. I made crumpets and tea and I was so grateful. As more people came, so too did the police. There were vans-full, on every corner and hiding on other streets. The barriers arrived and the pathways were blocked off. The police were unfriendly at first, not allowing any footfall along the paths. Most had been drafted in from Ecclesfield police station and hadn't even heard of the campaign. Some ended up being quite friendly, and ended up using my toilet and the backyard for cigarette breaks. One officer told me he would be doing the same as us if it were happening on his street. But he was maybe one good guy amongst the rude and bolshie ones. Many supporters were squashed onto my front garden. I was so relieved I wasn't alone and amazed how so many had come out in the cold to stand together. There were arguments with the police and Amey crew, Alison [Teal] especially striving to reason with the idiots. The one thing I truly regret is not standing under my tree. However, I had been warned several times by my manager that if I were arrested I would face instant dismissal, as I need an enhanced record check for my job. This was made very real when she turned up at my house that day to reiterate this. I had to go into my own house and discuss this with her! This was my annual leave! How could she just turn up like that?! It galls me that a photograph was taken of the many who came to support Lonesome George and I was inside having a lecture, so I'm not in the photo. The only upside of this, is that I felt I could no longer work with this company and I left, finding a much better job. The day we all stood, watching my tree being cut down so brutally (and with glee by certain arbs) remains with me today. I cried as people drifted away and sobbed later on my own. I honestly felt bereft, almost like I'd lost a friend, and I suppose I had. Lonesome George's trunk was left, standing like a beacon of all the spite and malice SCC and Amey had been throwing at us. There was a glimmer of hope the tree could survive, so a few stalwarts kept vigil in an attempt to keep him alive, standing under the trunk, Astrid, Paul [Brooke], Richard [Kedward], to name but a few. Amey tried daily. They told us we couldn't be here all the time, that they'd get him eventually, etc. Then one day he was left alone for twenty minutes. I got a call at work and a photo of him being loaded onto the back of a truck. Another terrible day. I'm still angry, still sad. I'm thankful that the felling has stopped, and hopefully it has stopped for good.

Simon Crump: I regret Lonesome George very much. Hindsight is a wonderful thing, but I reckon if we'd gone over the barriers in large enough numbers that day we could have saved that tree, but we didn't. Overall, I wish I'd done more throughout the whole campaign. I didn't do enough, not nearly enough. But like I say, hindsight is a wonderful thing.

Elizabeth: The campaign was, and is, made up of lots of different people. There were arguments, differences of opinions, egos clashing etc, but you know what, those people who were out on my street that day, all of them were amazing and still are. The whole tree-felling scandal will stay with me forever. It changed me. Maybe the one positive out of all this is the fact more people are now aware of the need for trees, street trees especially.

Simon Crump: By the end of the week we were cold, tired and dejected. It had been a long and punishing week and a lot of us felt that there was a real disconnect between those of us trying to make a difference and save trees at street level and the then STAG Steering Group's views about the effectiveness of NVDA. On that Saturday, I recall that we all met up in the Sheaf View pub. The overall feeling was that all we could do was stick at it and brace ourselves for more of the same in the coming week, and that if we couldn't prevent fellings we could at least minimise them by using delaying tactics. I remember feeling so dejected that I left the pub early with my pint unfinished on the table. And I never do that.

Paul Powlesland [campaign barrister]: I had been falling more and more in love with nature, and as a lawyer there was something I could do. I posted online asking, 'Is there a law group that gives time for free for trees?', but it seemed not. I was put in touch with STAG by Rob McBride and spoke to Chris [Rust] who said, 'Thank God, we've had a nightmare week. We know the police are wrong but we need a Barrister's opinion. I don't know how long we can keep going'. This was on Friday afternoon at the end of the Chippinghouse Road week and I was on my way back from Oxford County Court. I was going to Scotland that evening for a break and I took my law books with me! Usually, I feel the law is against us, but this time I looked into the Trade Union law and thought, 'What the fuck?', and then, 'They've completely misinterpreted this law!'. The piece wrote itself really. It was a bit of an open goal [for the Council]. I wrote the last bit on Saturday night while eating haggis and drinking whisky trying to get as much put together as possible! The immediate effect that it had almost never happens! To have that instant impact was the best thing I've ever done, my best work, it's the achievement I'm most proud of. I used my skills in combination with what the protesters have all done. My role would have been utterly pointless were it not for the people on the streets. My role was to allow the direct action to continue. It was weird that the very week I posted my query online was such an important one for the campaign, that week of all weeks! There was a moral clarity to this campaign and it has fundamentally changed my life. I'm so privileged to have been part of it! I really enjoyed the court cases too, even when we were up against the odds! There are thousands of lawyers who are happy to stand up for the people destroying our planet, but very few to stand up for those trying to stop it.

Calvin: The campaign was never the same after that week on Chippinghouse Road. There was a division throughout the week between two distinct courses of action. It was important for several reasons. It was the first big hit on Nether Edge, the test of our new tactics, the peak of the police use of 'TULCRA', and, most importantly, the week that the Sheffield tree campaign became much more than just a campaign group and became committed to the physical, peaceful defence of our streets. It was hard work, dispiriting at times, but with hindsight it was the birth of the campaigning tactics that were to succeed for us in the future. It was a sea change and where the battle for Sheffield's street trees really took off.

2017

'I want to know where you live. I'm going to follow you home.'

Resident (to Alison Teal)

St Ronan's Road is a tree-lined street in Nether Edge adjacent to a main road which is an air-pollution blackspot.

Bracken Moorland: I'd become aware of the tree issue during 2016 when I spoke to campaigners outside the Town Hall. There was a roadshow event going on inside with councillors, officers and Amey staff. As a Labour Party member I went in, and was told I could submit questions about what was proposed. I filled in several pages! It was the events on Rustlings Road and the combination of the tree-felling and policing there that turned me from supporter to activist in the campaign. When I was aware of the issue, but hadn't really become involved, I still thought if they came for the tree outside my house I would stand under it. Then one day, I saw barriers being set up at the top of the road and I went and stood under the tree which I can see from my front window. I called the Nether Edge tree group call-out number and was quickly joined by ten or fifteen people. That was the first time I was involved in street action over the trees. I'd been involved in various campaigns since I saved my pocket money to join the League Against Cruel Sports when I was eight. I can't stand destruction and I can't stand injustice. The ecology of the area was important to me as I remembered when, after a difficult birth, my world and capability had shrunk and my release was walking around Nether Edge pushing my baby son, walking around Montgomery Road, Kenwood Park Road and Chippinghouse Road. I'm not religious, but one day I was looking up at the trees in spring and it felt spiritual.

Alison Teal: My first experience of the trees on St Ronan's Road was when I was talking to a young mum on Steade Road whose house overlooked St Ronan's Road, and she was telling me about how important her view of the tree and nature was when she was at home with her baby.

Bracken: When they came to fell trees on Chippinghouse Road, just round the corner from me, that was very upsetting. I love drawing and painting those trees. The horse chestnut trees there are magnificent. I've never seen a blue whale, but I feel that those trees are the inner-city equivalent. The following week they were scheduled to come to my road and after nine people were arrested on Chippinghouse Road under TULCRA, the Trade Union law, I asked Inspector Stubbs, who was in charge of the operation, whether I would be arrested if I stood in my own garden, if I was preventing the felling of the tree outside. He said that their legal advice was that I would be breaking that law.

Calvin Payne: That Monday morning, we started to gather after the long, hard week on Chippinghouse Road and the police arrived and announced that they were withdrawing their support of the felling crews after Paul Powlesland's Barrister's Opinion about the use of the Trade Union law. That was a great relief.

***

Simon Crump: The next time we were on St Ronan's Road was on 23rd May 2017. It was the first day that they were gathering evidence for what we now know was going to be the High Court injunction. I'd been on Abbeyfield Road in Burngreave that morning and seen the 'evidence gatherers' in action.

Alison: That morning on St Ronan's Road it had felt quite hopeless, we didn't have enough people and there was hostility from some residents. The decision was made to withdraw. We had left and I was back at home making lunch. I began to think about the young mum I had talked to previously and I felt bad. The residents who didn't want us there had made us uncomfortable and I had a sense that we had abandoned the road and the people who opposed the felling. It was guilt really, it wasn't right that we'd allowed ourselves to be bullied by residents, so I went back and got bullied by residents!

Simon: I got back home from Abbeyfield Road, and saw on Facebook that Alison had gone back to St Ronan's, after what I already knew had happened in the morning there, because I'd been following it on our WhatsApp group. I saw that she was back there and my first thought was, 'Oh fuck, I'm going to have to play the shite in whining armour bit here'. I really didn't want to go, but even so, I went over. When I got there, as soon as I came round the corner, I was surrounded by a large group of very, very angry people. The next thing I remember was regaining consciousness on the floor with a policewoman kneeling next to me asking, 'Are you OK, love?'. So that was fun.

Alison: This was before the injunction, so I went into the work zone. Instantly, a group of angry men encircled me. I told them that I was a local councillor and that it wasn't a good idea to treat me like this. I pushed my way through them and made a run for it. I made the call-out to tell people I had returned to the road. The men were standing shoulder to shoulder outside the zone, to stop anyone else getting in to support me. It was in trying to fight through those people that Simon found himself on the floor. I called the police as I considered I had been assaulted and they attended. I wasn't hurt but I was very shocked. Other campaigners had started arriving by then. It was difficult, we were clearly not welcome.

David Glass: When I arrived, Alison was surrounded by angry people. She pointed out another tree that was under threat and so I went to stand under that one. As I went for a gap in the barriers, they were pushed into me. A resident tried to get in my way and pushed me. As he did so, I told him not to touch me and he started shouting 'Assault!' at me. Two police officers took him to one side and told him that he couldn't do that to me. He then made a complaint against me for assault. The police told me that I needed to come to the station and make a statement. I offered to do it later, as it would have meant me leaving the tree unattended. It was the first time I'd stood under a tree as I travel a lot and have seen what having an arrest on your record can mean for people visiting or for working in America, for example. I found myself accused of assault, when I had earlier been roughed up myself. It was also nesting season and I was asking, 'Have you inspected for nests?'. What I was doing could have been preventing a crime.

Bracken: I saw the crews on the road, went and stood in the barriers, and a resident shouted at me whilst another barged me out of the way. The man shouted, 'Get out of the way you stupid woman, let them get on with their work! Why are you stopping them from doing their job?'. These were people I had seen and spoken to for the thirty years I've lived here.

David: The accusation of assault turned out not to be about the disagreement we'd had but an entirely different made-up story. He had said that I'd pushed an old lady over. Three people made near-identical statements making this allegation which was nonsense. It hung over me for three weeks and then was dismissed for 'lack of evidence'.

Calvin: They always say that when it didn't happen in the first place.

Simon: It got nastier and nastier, and a woman who witnessed me being threatened was told by one of the people concerned, 'If you make a statement, we know where you live.' It was very unpleasant.

Calvin: We all left together, pursued by some of the angrier people on the street. A documentary filmmaker that was with us got into an altercation and we had to get him out of there too.

Alison: A man followed us, filming us as we began to leave. He said, 'I want to know where you live. I'm going to walk home with you'. I said, 'You can't, I'm in my car!'. It was a difficult day. It wasn't anything heroic to go back that day, it just felt like something that I needed to do. I just felt that you shouldn't let bullies win.

David: I was told that someone on the street had poisoned a neighbour's cat before. I changed my Facebook profile after that as it used to have my dog in the picture and I didn't want to be identified that way.

Alison: I think this was the first time a picture of me inside a work zone had been posted online. It was this that led to Sheffield City Council accusing me of bringing the Council into disrepute under their code of conduct. That went on for well over a year.

Simon: The next hour was probably the most frightening, for me, of my experiences during the entire campaign. The residents were just furious, it was almost as if somebody had primed them, gee'd them up. It seemed that every single thing which had ever gone wrong in their lives was a direct result of the trees on their street. I eventually managed to get under a tree, just as David Glass was being carted off. I replaced him, and a reassuringly big copper came and stood with me inside the barriers, 'For your own safety', he said.

David: When I made a complaint about what had been done to me on the street I was told that, 'If you go on a protest you've got to expect a bit of push and shove'. There appeared to be a double standard.

Simon: I think the police were genuinely shocked by the level of violence and aggression that day. Later on that week, I was standing on a street corner waiting for Calvin to do an interview with ITV *Calendar News* when a police car pulled up next to me. It was the WPC and the big copper from the day before. 'What are you up to today, Simon?' she asked. 'Oh, I've just gone back to selling drugs', I said. 'It's a lot more fun than what fucking happened yesterday'. 'Much safer', she said, 'If you get in any trouble you know where to find us!'

Bracken: Was it worth spoiling the community cohesion for the tree-felling? That will take a long time to repair. There are people I no longer speak to, people I've known for many years. Neighbours that I used to smile at and say 'Hello' to. I don't anymore, and I'm sure that's mutual.

Calvin: The last time I was on St Ronan's Road was at around 4:30 one winter morning after a night raid. I didn't do much of the night patrolling duties, but on this occasion I was with Susan [Ashworth] driving around looking for Amey crews. By then they were desperate enough to be out chopping off branches in the early hours so that it was quicker and easier to fell the trees during the day. We went past St Ronan's Road around 4am, returning half an hour later to find a lot of work had been done. The tree at the top of the road had been stripped of many of its branches.

Bracken: On the positive side the 'STARTS' art group came out of one of my lowest days in the campaign. The first injunction trial was starting in Leeds and I went and painted a picture of the magnificent horse chestnut trees on Chippinghouse Road because I thought we would lose them. I did something I had never done before and lay on the ground looking up at the canopy. While I was lying there Sarah [Deakin] came around the corner and asked what I was doing! From there we organised an art event, then another, and it grew into something important to the campaign. I'm immensely proud of what we did, so pleased to have met so many great people. We need an inquiry into all the unexplained decisions and actions. Why should Paul Billington get to retire early and be paid off? Why should the police get away with what they did? Why are there so many unanswered questions? I'm still angry and want those answers, especially with similar protests going on over HS2 and others. There needs to be some answers.

Calvin: Looking back, there are so many mixed feelings. The achievement in winning along with the anger and frustration at the things we endured along the way, not to mention the damaged relationships on many streets.

Bracken: I've been involved in many campaigns and causes in my life. The big difference with the tree campaign was that it took over. It didn't end when you went home, when you shut the door. It wasn't something you could pop in and out of. It took over every day, it was on our doorsteps, every morning, every evening, every night. It was all-encompassing and exhausting.

THE INJUNCTION

General Form of Judgment or Order	In the High Court of Justice Queen's Bench Division Sheffield District Registry	
	Claim Number	D92LS739
	Date	20 July 2018 2 0 JUL 2018

SHEFFIELD CITY COUNCIL	1st Claimant Ref	
ALICE FAIRHALL	1st Defendant Ref HR/BM/RY/DILLNER/11744	
SIMON CRUMP	2nd Defendant Ref	
REBECCA HAMMOND	3rd Defendant Ref	
ALISON TEAL	4th Defendant Ref	
DAVID DILNER	5th Defendant Ref	
CALVIN PAYNE	6th Defendant Ref	
PAUL BROOKE	7th Defendant Ref	
GRAHAM TURNBULL	8th Defendant Ref	
ROBIN RIDLEY	9th Defendant Ref	
PERSONS UNKNOWN	10th Defendant Ref	

10th Defendant - PERSONS UNKNOWN BEING (I) PERSONS ENTERING ANY SAFETY ZONE ERECTED AROUND A TREE AND/OR (II) DELIBERATELY SEEKING TO PREVENT THE ERECTION OF OR TO INTERFERE PHYSICALLY WITH OR WITH THE USE OF ANY SAFETY ZONE AROUND A TREE AND/ OR (III) REMAINING WITHIN ANY SAFETY ZONE AFTER IT IS ERECTED AND/OR (IV) KNOWINGLY LEAVING ANY VEHICLE IN ANY SAFETY ZONE OR INTENTIONALLY PLACING A VEHICLE IN A POSITION SO AS TO PREVENT THE ERECTION OF A SAFETY ZONE AND/OR (V) PREVENTING DELAYING OR SLOWING DOWN (FOR MORE THAN 20 MINUTES IN ANY 24 HOURS (WHETHER INDIVIDUALLY OR AS PART OF ANY GROUP)) ANY CONTRACTORS (ENGAGED IN ACCESSING, EGRESSING OR CREATING ANY SAFETY ZONE) IN THEIR USE OF ANY PUBLIC HIGHWAY WHICH IS THE SUBJECT OF A ROAD CLOSURE IN CONNECTION WITH TREE WORKS WITHIN THE ADMINISTRATIVE AREA OF THE CITY OF SHEFFIELD AND/OR (VI) ENCOURAGING AIDING COUNSELLING DIRECTING OR FACILITATING ANYBODY ELSE TO ANY OF THE MATTERS ABOVE

The court office at Sheffield District Registry, The Law Courts, 50 West Bar, Sheffield, S3 8PH. When corresponding with the court, please address forms or letters to the Court Manager and quote the claim number. Tel: 01142812400 Fax: 01264 347961. Check if you can issue your claim online. It will save you time and money. Go to www.moneyclaim.gov.uk to find out more.

Produced by:Imran A
N24 General Form of Judgment or Order
CJR065C

INCLUDING BY POSTING SOCIAL MEDIA MESSAGES WITHIN THE ADMINISTRATIVE AREA OF
THE CITY OF SHEFFIELD

TO THE DEFENDANTS

PENAL NOTICE

IF YOU DO NOT COMPLY WITH THIS ORDER YOU MAY BE HELD TO BE IN CONTEMPT OF
COURT AND IMPRISONED OR FINED OR YOUR ASSETS MAY BE SEIZED

AMENDED ORDER

Before His Honour Judge Robinson sitting as a Judge of the High Court on 11-12 July 2018 at the Law Courts,
50 West Bar, Sheffield, S3 8PH

UPON hearing Leading Counsel and Junior Counsel for the Claimant.

AND UPON the Tenth Defendant being unrepresented

AND UPON the Court having made an order on 18 August 2017 for injunctive relief against the named Defendants
which expires on 25 July 2018

AND UPON the Court having made an order on 18 August 2017 for the same injunctive relief against Persons
Unknown which also expires on 25 July 2018

AND UPON the Claimant having issued an application notice dated 28 June 2018 for (amongst other things) for
continuation of the said injunctive relief after 25 July 2018

IT IS HEREBY ORDERED

1. The Tenth Defendant must not from 26 July 2018 until 23:59 on 25 January 2020:
    a. Enter any safety zone erected around a tree; and/or

    b. Deliberately seek: (i) to prevent the erection of; or, (ii) to interfere physically with or with the use of any
    safety zone around a tree (which prohibition shall (without limitation) include the following acts:

    • Moving, lifting, pushing, pulling, damaging or destroying any features demarcating the safety zone or
    climbing upon such features;

Sheffield City Council
Culture & Environment
Sheffield City Council
Howden House
Union Street
Sheffield
S1 2SH

Sheffield City boundary shown red

Sheffield City Council
Culture & Environment
Sheffield City Council
Howden House
Union Street
Sheffield
S1 2SH

Sheffield City boundary shown red

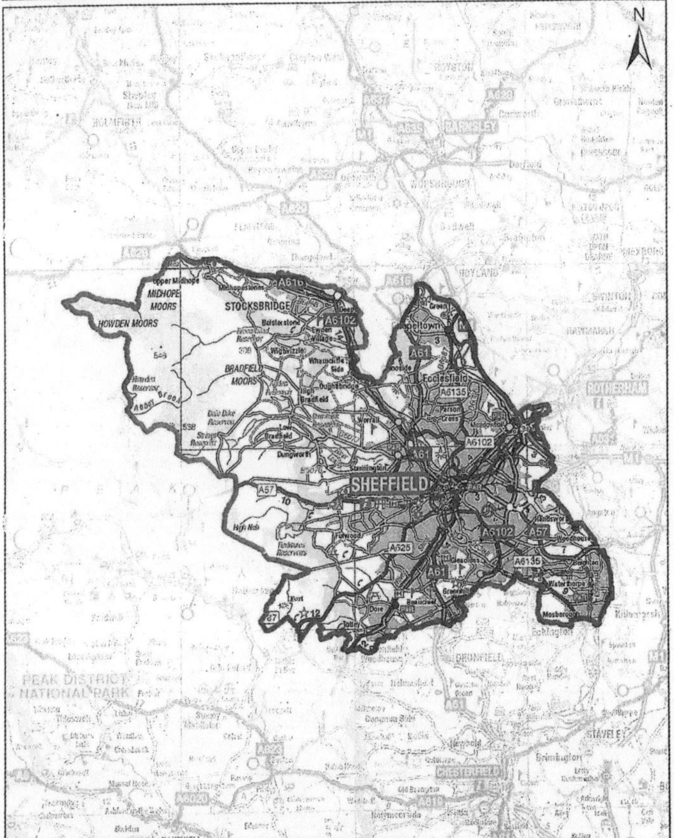

23rd August 2017

'...a motley crew wearing dressing gowns, wigs, a paper maché Paul Brooke head, and other assorted fancy dress...'

Mark, Protester

Olive Grove Road is a partially-terraced street in the Heeley area. It houses the Amey 'Streets Ahead' depot. It doesn't have any street trees.

Mark: Sadly, the Council gained their injunction. The judge's refusal to look at the contract and his comments about Paul Billington being an 'honest and reliable witness' gave the impression that we were battling Establishment bias as much as points of law in these situations. The Council thought they'd won, and for us the question was now, 'How do we stop them felling trees when they can take us to court if we enter the 'safety zone' around a tree?'. Various ideas about masking up to conceal identity were being suggested, but a few of us wanted to make a big statement that we weren't giving up. If we couldn't stop fellings 'on site' then what about stopping the crews getting to the trees? We needed to stop them leaving the depot! Amey's depot was called Olive Grove, even though we'd never seen any olive trees, and the reality was that it was a pretty grim site, the air thick with exhaust fumes from the constant vehicle movements in and out. Paul [Brooke] suggested that 'slow-walking' was a form of direct action that wouldn't result in arrest for obstruction, so we began planning a slow-walking 'crocodile' of people that would just happen to want to walk the pavement outside the main exits of the depot. There were a few problems though, Acorn [the main tree-felling subcontractor] had their depot in Darnall, a couple of miles away, and if they and Amey got wind of our plans they might just station the arb vehicles elsewhere and go directly to the trees, therefore bypassing us. We also needed enough protesters to make the blockade work. A few early mornings were spent 'staking out' Olive Grove and the Acorn depot to try and get a picture of the routine and 'Theresa Green' posted ideas on Facebook about a protest to happen on the first day the injunction came into force. The idea was to get people to save the date without giving too much away, we even tried to wrong-foot the Council into thinking the protest would be in a felling area like Nether Edge. We knew the Council were monitoring the STAG Facebook page from the evidence used in the hearing. I still wonder who had the job, paid for by our Council Tax, of monitoring Facebook and Twitter accounts to feed information to Council officers and Amey managers.

Simon Crump: The fact that the protest took the good-natured but defiant form it eventually did is all down to Mark. I was all for high drama, hysterical outbursts and locking on and getting arrested like I'd seen the anti-frackers do, which of course would have completely ruined the party atmosphere. Mark definitely had the right idea here, and to his great credit, he managed to hit the same cheerfully defiant tone again when we occupied the Council chamber.

Mark: The core group 'in the know' about the real plan was kept to a handful of trusted people. It seemed that the Acorn crews always called in to Olive Grove each morning to get their job lists, so with careful timing, and a bit of luck, we hoped we might just be able to trap all the crews. The 17th of August dawned and a motley crew wearing dressing gowns, wigs, a paper maché Paul Brooke head and other assorted fancy dress arrived on the neighbouring street. The fancy dress was a regular theme in the protests, a counterpoint to the aggressive hostility of the Council and sometimes a way to diffuse tense situations on the streets. We gathered at the depot entrance in case Amey arbs tried to leave, but we didn't want to make too big a scene in case the Acorn crews sussed us out and got diverted. I was on tenterhooks until we watched the Acorn vehicles drive in. They 'clocked' us, but obviously hadn't realised what was happening – the trap was sprung! The alert then went out via WhatsApp groups and Facebook that the protest was at Olive Grove, so our numbers swelled to a jovial crowd. The 'slow walk' began, but we let the non-arb vehicles out, as we thought stopping works like pothole repairs from taking place wouldn't be popular with the public. This was fine until they tried to get the arb vehicles out and created a traffic jam. We held firm though, and the arbs did some manoeuvring and went back to the canteen. Amey managers came out to reason with us, then to threaten us with the police. The police came and found no crime being committed, much to the chagrin of Head of Highways, Phil Beecroft. We later found out through an FOI request that he had been emailing the police demanding that they arrest the perpetrators!

Simon: At the height of the campaign, I would get a call from Josh at the *Guardian* (those were the days eh?) every Sunday, asking me if we had anything planned for the coming week. And I could never tell him anything because we were all over the city. But when I got the call on the Sunday after the injunction was granted, it was a very different story. 'Anything planned for this week Simon?' 'Well...can you keep a secret?' I said. And then I told him what we were going to do. Josh was there very early that morning with a photographer [Chris Thomond], even before most of us were, and he did us proud, staying for hours and then writing a really brilliant piece for us.

Mark: Amey then sent private detective Richard Wood's 'process servers' out to us, 'Boris' (who looked a bit like Boris Becker) covertly (but not that covertly) filming us from his car, and the 'Chuckle Brothers' waving injunction paperwork at us – we weren't under a tree so it didn't apply! We decided to make a tactical withdrawal at the end of the day – the Council had thought their expensive injunction would stop the protests – but we'd stopped them felling any trees, got some great media coverage and, most importantly, showed them that we weren't beaten and still had plenty of ideas and enthusiasm up our dressing-gown sleeves!

Calvin Payne: I hadn't been involved in the planning for the event that day. It was a day of personal mixed feelings. I had the impression it was a 'last hurrah' for some, the last chance to cause some frustration for the felling crews. I was saying that I was going to carry on regardless of the injunction, and wasn't getting any backing. I was treated dismissively that day for saying we needed to be prepared to breach the injunction, and for a few weeks afterwards too, but I probably didn't endear myself to people sometimes. I was on my own for a short time, while also summoning up my own courage for what was to come. Most importantly though it was a great day for the campaign from start to finish, totally taking Amey by surprise and giving momentum to our side just at the moment the Council and Amey thought they had achieved their victory.

Jenny Saul: When my son Theo found out that the tree outside our house was marked down for felling, he was really upset. He said that he would do whatever he had to do to save it, adding, 'I will die for that tree!'. I explained to him that although there were causes which were worth dying for, this was probably not one of them, and although he should fight for it, dying for it would be going too far! The depot blockade came around during the school holidays. Theo really wanted to get involved so it was the perfect opportunity for him. There had been a suggestion that we all attended as 'Persons Unknown' in disguise and Theo already had a great interest in disguises, in fact he'd asked for 'disguises' for Christmas the previous year. So he went off upstairs to try out various things and came back down wearing a red silk Chinese robe and a NASA space helmet, which was a really fabulous-looking outfit. So the call-out went out and we were there in disguise, Theo in his outfit and me in a knitted octopus mask! The disguise element made it a really fun event for kids and Theo had a great time slow-walking across the entrance to the depot. Lunchtime came around, and while we were waiting for our lunch, that picture came up on the *Guardian* page, and it was me and Theo with Simon [Crump] and Susan [Ashworth] in the background!

What he really liked about it was the fact that he could tell people about it, but that it was his choice whether or not to say it was him!

Martin Pickles: In the days preceding our assembly at Olive Grove Depot, there'd been talk on the WhatsApp group about a bit of fancy dress. Dressing gowns were easily chosen in honour of the wonderful women of Rustlings Road who'd stood up to the felling crew and police at 5am back on 17th November 2016. This protest on the first day of the injunction would see lots of slow-walking across the entrance of the depot to prevent barrier lorries and chippers leaving. I arrived bearded in a blue paisley dressing gown and Farrah Fawcett wig (thanks to my wife's disco classes). Frankly, it was a disturbing sight. I would like to take this opportunity to apologise for any distress or trauma I may have caused. However, I maintain what I posted on Facebook at the time: 'I may look ridiculous but I'll never be as ridiculous as Sheffield City Council!'.

Sarah Deakin: It was 23rd August 2017, and I'd joined the fancy dress protest of around fifty people outside the Streets Ahead depot at Olive Grove. This was the morning that the High Court injunction came into force, obtained by Sheffield City Council to prevent tree protectors from taking 'unlawful direct action' to protect street trees. Protesters were no longer able to stand inside barriers to prevent healthy street trees being felled for fear of prosecution, and no doubt SCC and Amey thought this landmark ruling meant they'd won and we'd now go away. What the injunction didn't do was prevent other forms of protest. So, if we couldn't directly protect trees where they stood, what could we do instead? This was a great gift to the creativity of the campaign and plans were made to frustrate the workers getting to the trees. I wore my pyjamas and my other half's dressing gown, complemented by a blonde wig, false nose (complete with huge boil) and sunglasses, plus a 'Persons Unknown' sign slung round my neck like a prisoner's ID tag. The injunction named individual defendants plus 'Persons Unknown', which effectively meant anyone in the world who tried to defend a street tree by getting inside the work zone. We celebrated this new moniker by brandishing signs and placards bearing 'Person Unknown' in an 'I'm Spartacus!' style. At the depot entrance, as ever, we were entirely peaceful and good-humoured, milling about, displaying our banners to passing traffic and getting lots of beeps of support. Every time an Amey felling crew tried to exit the depot, we walked slowly back and forth across the exit, exercising our right of way as pedestrians and not in any way blocking a highway or blockading the exit. After making a few attempts to leave, this obviously became highly frustrating to the crews as there was nothing they could do. They tried to leave by another exit, not realising that there was a second contingent of protesters there too, and gradually it began to sink in that perhaps their day was scuppered. I remember this as the first day I'd witnessed Amey 'evidence gathering'. Two guys who would become well known to us came out of the depot to read us a statement. The reader was a short fella with dark hair and glasses, later immortalised in a cartoon by Lydia Monks, and the other a big blonde chap who we later nicknamed 'Boris' (after Boris Becker). I can't remember the details of the statement, but the gist was we needed to leave or they'd call the police. All the time this was being read out, 'Boris' was filming us on his phone, and I remember a stand-off between Dave Dillner and him as they stood face to face filming each other. We'd see both evidence gatherers many, many, times after that, including at some of the notorious protests such as Millhouses Lane and Meersbrook Park Road. Of course the protesters didn't move, so they called the police who came, but couldn't do anything as no laws were being broken. I had left by this point but shortly after word arrived that Amey had cancelled their work for the day. So much for their groundbreaking injunction! That day sticks in my memory as one of the most fun and creative of all the '#SaveSheffTrees' protests. I met loads of people that day that I'd only ever talked to online before. It was hilarious trying to put disguised faces to names and even funnier in the weeks afterwards, when we had to do the same again without masks, wigs and pyjamas! The protest gained huge press coverage, with journalists from BBC, ITV, as well as local and national newspapers. The highlight for me was being in the *Guardian* 'Photos of the Week'. Who else can say their first appearance in the national press was in a photo showing them bedecked with a wig, false nose, sunglasses and dressing gown, along with some of the biggest rabble rousers in the campaign, and my dad still has no idea what I did that day!

September 2017

'You're probably one of those people who's opposed to fracking as well aren't you?'

Resident to Sally Anne Wickenden

Woodstock Road is a tree-lined residential road near a very busy air-pollution blackspot.

Sally Anne Wickenden: When we knew the survey letters for the Independent Tree Panel were due to arrive, I knocked on all the doors on the road. We had a lot of support on the street and I wanted to make sure that people replied. When the street voted to go to the panel, the result was that the decisions were re-examined and the number of condemned trees fell from fourteen to nine. There had been attempts before to fell a tree just around the corner from the bottom of the road outside Common Ground Community Centre, but each time people had been able to occupy the bank of land close to the tree and prevent the work. So when the notices were delivered for the week-long work we were ready.

Jeremy Peace: I live within sight of the tree outside Common Ground. On a couple of occasions before that week, felling crews had made early morning attempts to fell that tree. It seemed like opportunistic action on their part, as it is a single tree which isn't really on a particular road. It's next to the roundabout at the bottom of Woodstock Road.

Calvin Payne: The couple of attempts made on that tree took place before the High Court injunction was awarded in July 2017. It was at a time when preventing the felling was reasonably simple, as the police weren't getting involved. There was a build-up of two or three months before the case when the Council were waiting to see what happened in court.

Jeremy: It was a bit strange sometimes. Attempts at felling seemed half-hearted – they didn't seem too miffed to see us there.

Calvin: The Woodstock Road work was scheduled just a couple of weeks after the injunction came into being. It was the first test of both side's resolve and what happened here was going to affect what happened afterwards. We set out to ensure that we had as many people as possible on the road, being as awkward as possible. Whatever else happened we were determined to make them work to the law, we wanted to see Health and Safety paperwork and proof of wildlife inspections. Things they had a legal obligation to carry out.

Sally Anne: We didn't know what to do post-injunction. We did think that signs in windows refusing permission for the arbs to oversail private property would work at my end of the road, where the trees are closest to the houses. Over the weekend before the work was due to start I was knocking on doors and I spoke to a new tenant who was keen to prevent the felling. I wanted him to take one of the 'No Oversailing' signs for the window. He happened to be on the phone to his landlord at the time and he passed the phone to me and said 'You speak to him'. The landlord told me that he wanted the tree felled as it was, 'dying and is going to fall on the house', which wasn't correct. He said to me, 'You're probably one of those people who's opposed to fracking as well aren't you?'. To which I replied 'Well, yes, I am!'. I persuaded him that his tenant wanted to keep the tree as did most of the people on the road and there was no real reason to fell it. It wasn't dying and should be saved. He listened and didn't argue any further. We got our garden permission.

Calvin: So we spent the week occupying the gardens where we had permission to do that and making a nuisance of ourselves on the street. There were residents on the street on both sides of the argument (as was the case on almost every street in the city). If people believed the Council propaganda that the trees were diseased or dying, all we could do was to give them the Council's own statistics which proved that wasn't the case. The work zones were made large enough to try and keep us away from the trees and to cut off access to the properties where we had permission to be. I carried a list of the garden permissions and knew that one of the addresses right next to a particular threatened tree was a campaign supporter's home. The lead arb refused me permission to access the garden, even though I had permission to do just that from the owner of the house. Instead, I had to climb over several (admittedly very low) neighbouring walls to access the property. When I got into position, the tenant from the house came out to see what was going on. I explained, and they got the owner (whose name I had on my list) on the phone. There followed a bizarre conversation which went, 'Hello, you don't know me but my name's Calvin from Save Nether Edge Trees and I'm stopping your tree being cut down.' To which he replied, 'That's great, but I'm in Hungary!'. The arbs and the police never seemed to quite believe the reach of our garden permission efforts, but this was an impressive one even for us.

Jeremy: On the same day a campaign member who is a university professor needed to deliver papers to a colleague whose house found itself on the wrong side of the work zone. She said that she had to deliver them in person, as they were important and confidential but the arbs said that they would take them for her, which she refused [to allow]. As we were becoming more reliant on garden permissions, the other side tried increasingly desperately to stop us accessing the properties in the first place.

Calvin: There was a fair amount of abuse aimed at us that day. Residents would often accuse protesters of not being local, or failing that, not living on that street and in one case, not in that particular house! Whereas some of us may not have lived on that street, all of us lived in the same city. We all had an interest in the wellbeing of Sheffield and our community. I was told by a resident later that things got 'a bit awkward for me in the area for a while' once they became associated with the protests. Communities and streets were divided in a way that we felt was being encouraged by the Council and Amey. At the end of the week, just two of the nine scheduled fellings had taken place. I was asked more than once how we had prevented them, when we were no longer entering the safety zones. The fact is I didn't really have a simple answer. We were slowing the arbs down as much as possible, challenging everything. Most of all, though, we continued to turn out in large numbers. I always said that if we turn out, we don't know what will happen, but if we don't turn out then we know only too well.

September 2017

'I remember the arbs gave out safety masks to the residents nearby who were on their side, but not to us who were trying to protect the tree.'

Isela Gonzalez Vazquez, Resident

Sheldon Road is a tree-lined residential road in Sheffield near a very busy air-pollution blackspot.

Calvin Payne: Coming so soon after the High Court injunction, garden permissions were paramount. The street trees were large enough and close enough to the front gardens to be protected that way where residents were supportive. On Monday morning the barrier guys were setting up around a large tree on the lower part of the road and were intending to close off that part of the street. I watched from the other side of the road for quite a while and hadn't been recognised. As they finished setting up one of them came over and said, 'Sorry mate, I'm going to have to ask you to move in a minute. We're closing this part of the road.' To which I replied, 'In a minute I'm going to be coming over and standing in that garden to stop the tree being felled'.

Isela Gonzalez Vazquez: I was just leaving my house, I had my lunch packed and everything, ready to head off to university. As I went out, I was trying to remember who had given me the emergency number in case we saw Amey crews on the street. I thought, 'Right', as I walked out of my flat and then saw this going on and I found the call-out number and called it. I thought I might as well walk down there and see what's going on and then I saw all these wonderful people there. I think that was the first point of contact for me, and then I just stayed the rest of the day. With my packed lunch.

Calvin: One of the trees that we had a garden permission for, at the bottom of the road nearest the main road, was lost when the neighbouring house persuaded our supportive resident that the tree was damaging their driveway. Neighbours were often divided and there was a lot of pressure on people, with the Council's claims which made some of them afraid of damage or facing future problems with insurance. Giving a garden permission often meant several protesters spending much of the day on your property, and when a street was divided that was a brave and principled permission to give. The next day that became more stark when work was begun on a large plane tree further up the road which had three houses within its canopy. The first house under the tree had a group of young male tenants who spoke little English, the second had a very pro-felling woman owner, the third a family of campaign supporters. As any one of those houses was close enough to prevent the felling work to progress much further, we tried to work out the way forward within the terms of the new injunction.

Jeremy Peace: We were trying to get our heads round the terms of the injunction but so were they (Amey).

Calvin: It was standing there that I found out just how much fine green dust a plane tree can give off. It was choking and got everywhere.

Isela: I remember the arbs gave out safety masks to the residents nearby who were on their side, but not to us who were trying to protect the tree. So when we were trying to think of a strategy, what we came up with I remember, was that I, as a neighbour who lived across the street, would knock on their door, ask for permission to go through the barriers and then, because they weren't providing anyone with masks at that point and there was a lot of dust and they had the window open, I would go knock on the door and say, 'You might want to close your window because there's a lot of this dust and it's bad for you'. And then take the opportunity to sort of talk to them about what was going on. And so that's what I did, ask for permission; they actually let me right through.

Calvin: This was the first time since the injunction that I saw the importance of being a 'Person Unknown'. There was definitely a bit of leeway given to Isela for not being known to the arbs when trying to access a work zone or property, compared to what would have happened if I, or someone else from the court case, had asked the same thing.

Isela: I said, 'My name is Isela, I live across the street, nice to meet you, thank you'. Then the resident signalled to everyone who was standing across the street that it was fine to come and stand in their garden. We were all standing there, and I think Amey staff were walking around with a piece of paper saying that they had been asking for people's permission, and actually there was something on there about not only allowing Amey to work on private property but also not allowing protesters in the garden. And so there was this guy walking around with this list and he said, 'Well actually I have permission from this property and it says this, it says you can't stand here and we have permission to work here'. So then a few seconds later, the next-door neighbour, a woman, very aggressive, walks out and demands that we leave, saying that we have no permission to be there.

Calvin: She wanted to control what her neighbours thought and did. It got a bit 'push and shove' for a moment. I don't know what the issue really was, sometimes tempers were lost for unclear reasons.

Isela: So we had permission, but she was saying, 'You have no permission'. She's screaming at us at this point, 'You have no permission to stand there! Leave or I'm going to call the police' and I was like, 'Actually we do have permission' and she was like, 'No, no, no, no, you knock on his door and get him out because I need to talk to him', and I'm saying, 'You knock on his door', you know, like we have permission. So she walked out, went through the back door, we could hear her banging on the door. She almost dragged him out, the poor guy was a young guy, like early twenties. Yeah, and he doesn't understand English, and I know this because I had just had the previous interaction. So she drags him out and the Amey person is joining them at this point holding the paper with the signatures, and makes him sign the thing, the piece of paper that says we didn't have permission to be there, and the guy just sort of looks at me with an expression that said 'Sorry'. Of course after my first interaction with him, I thought, 'What's going on?'. And so at that point she's screaming at us again, 'Get out or I'll get you out!'. We were, like, fine, we had seen what had happened, we were making our way out and I was the last person towards the end walking out of the garden. We were walking out, we weren't running out, and then she was still there. I can feel her right behind me, 'Get out or I'll get you out!'. I said, 'Listen, we're going in a bit'. I felt like she was going to push me, so I sort of turned around and I said, 'Don't you touch me!'. And she's just sort of in my face shouting. I'm saying, 'Do you see us, we are walking out of here'.

Calvin: It was around that time that Amey paid our garden permission tactic the compliment of copying it. Previously, residents who supported the felling allowed barriers to be set up in their gardens and driveways, but it was done on an ad-hoc basis. Now the residents were given pieces of paper to sign which gave Amey crews permission to set up on people's property as well as keeping us off. They weren't so keen on the legalities when working on trees that were oversailing properties where the resident was on our side.

Isela: I think I was very surprised to find out about a lot of my neighbours that I didn't know, then to sort of get to know them as a potential enemy for the rest of the time that I lived there. It is a strange feeling, like I shouldn't say 'Hello' to them when I walk past, and it's not a very nice feeling to not have a good sense of community, and also [to know] that there are people that want trees gone, for no good reason. And not only that, but also people who are prepared to be aggressive and abusive to achieve it.

Calvin: That was the week that the BBC came to Sheffield to record a piece for *Countryfile* about the tree-felling. They spent a couple of days on Sheldon Road talking to people on both sides and getting footage of the felling crews in action.

Jeremy: When the cameras were there the arbs were lowering down small pieces of tree with ropes, something they wouldn't normally be doing. They were being delayed by our close proximity and the presence of the BBC.

Isela: So during that week, my cousin was in town visiting and it was kinda bizarre, I guess, trying to explain it to somebody else. I said, 'Oh, I'm going to be outside', so well, I thought, she was on holiday so she's entitled to sleep in. I guess the weirdness of it all sort of cropped up.

Jeremy: Tell her this is how we spend our leisure time in the UK!

Isela: She was asking questions like, 'If everyone is opposed to it why are they cutting the trees anyway? Why is all the felling going on?'. And all the obvious questions, like had people gone to the Council to talk about this? I tell her, 'They've gone many times, they're just not listening'. So there were just a lot of questions about what's happening, does it make any sort of sense? I said, 'Nope, it really doesn't make sense'.

Calvin: There was a lot of art and performance in the campaign, and most of it was all new to me! One very impressive piece was when Katrina [Nice] made a stag out of small branches and twigs from a tree on Sheldon Road. It was so well done and accompanied a few of us to a social gathering in a nearby pub. It survived for some time in a garden afterwards too. We were a creative lot!

2017

'I don't know what else I can do, but I know I can do this. No-one's done anything like this yet.'

Tim Smedley, Resident

Dunkeld Road is a residential road in the Millhouses area of the city. It is adjacent to one of the busiest roads in Sheffield.

Jane Avgousti: Until Rustlings Road, I was prepared to give the Council the benefit of the doubt. I started taking notice of which trees had the felling notices on them and which hadn't, and I couldn't understand seeing beautiful, healthy trees in their prime condemned. It was a shock when it came to our road, and we saw it all for ourselves. Thinking back to the Council survey, it seemed to suggest that only dead or diseased trees would be removed which no-one disagreed with and that caused the idea that we were unreasonable. When I spoke to a few of my neighbours they were in favour of the plans and I wondered, 'Am I missing something here?'. However just because I support removing those genuine cases doesn't mean I was going to go along with deforestation of the city.

Tim Smedley: I'd thought the situation was bonkers early on and then Rustlings Road infuriated me. Total madness. At the time we were in the process of moving to Millhouses, well aware that the tree outside our new house was for the chop as was another nearby.

Rachel Rea: After Rustlings Road I was keeping a tentative eye on what was going on. I had been outraged by the use of Trade Union law there, and then one day realised that a tree near us was threatened. I looked at the trees listed for felling on the road and couldn't understand the rhyme or reason for it. I just thought that there was something not right. It was then, during that first couple of days, that we first saw that there were people from across the city turning out to stop the work. I joined in when I first saw the direct action, until somebody told me to watch out as Social Services had been called on protesters before, and I had my daughter with me.

Calvin Payne: This was January 2017. I came over and met a few of the local Millhouses tree group for the first time that week. Just a couple of months after Rustlings Road, so right in the middle of events.

Jane: I don't remember any notices going up or any warning. The police were there, not in the same sort of numbers as more recently, but it was still a shock.

Tim: The felling crew were just setting up. It was the first time they had come to the road and I didn't know what to do. I called my wife, Rachel, and told her what was happening and asked if she could call someone. She put a message on Facebook. Within minutes there were people here and we had enough to hold up the crews. We managed to stop them on Monday. I'm not sure how.

Calvin: I heard that the police were called away to an accident and the felling crews didn't want to proceed without them.

Tim: When they came back the next morning the police gave us half an hour to leave the safety zone when we tried to repeat our protest. They told us, 'If anyone is still in this zone after half an hour you'll be arrested'.

During that time I'd decided I was going to play something. I ran home and got my cello whilst thinking about what I could possibly play. By the time I got back under the tree there were only five minutes of the police warning time left. It was an opportunity for me to do something that hadn't been done before. If I hadn't done that, it would have felt like missing the one thing I could possibly have done, and I would have regretted not doing it, especially as it was on our road. My frustration looking back is that in the panic to set up I forgot to say the name of the piece, which is 'The Song of the Birds', a beautiful Spanish folk song. I'm really glad I did it. It attracted attention and I was interviewed by national and international radio stations as a result. It will always be a special moment for me.

Calvin: Later on the campaign it seemed that everyone was playing musical instruments, but not nearly as well as Tim!

Jane: I remember some of the residents who supported the felling coming out by that time. One of the repeated arguments we got was, 'I've lived on this road longer than you have'.

Tim: When the road was officially closed some of the residents held an 'anti-tree campaign protest' at the top of the street. It was a bit unpleasant having protests directed at us. We were protesting against the Council, not at any particular residents.

Jane: Looking back now I did a rather ill-advised interview with *The Star* talking about my son having asthma, which led to me being a bit of a target for some residents. The usual 'You're not from around here' argument caused some upset with my neighbour, who is Indian. She thought it was meant in a wider sense initially, but it was being said to a few of us of course.

Rachel: I remember when Susan [Ashworth] from the campaign was doing the bird survey on the street, and one of the vocal residents came out and, thinking it was about trees, said, 'We don't want any more trees on this street'. One or two people thought they spoke for us all on this road. It's been difficult to stay calm sometimes.

***

Jane: Over that spring and summer they came a few times and did some pruning. It did seem like they had us under surveillance, but nothing much happened.

Tim: We went away in August just as the injunction was coming in. We didn't leave our car outside as they could have put signs up and said we were breaching the rules. We expected to come back to the trees being gone.

Calvin: The parking services people did start being aggressive that week. Known supporters of the campaign had their cars towed and appeared to be treated differently from others. They thought the injunction permitted this. A woman from Amey knocked on one resident's door and told them they were breaching the injunction by having their car parked outside and threatened them with the full strength of the law if they didn't move it. One of our people filmed the exchange but unfortunately I was so angry at what was happening that I was shouting and swearing over the conversation. It would have been a very useful piece of footage and I ruined it.

Jane: Once they got the injunction they thought it was a done deal. They were frustrated when it didn't happen like that. Then there were weeks of relentless car parking and road closure notices going up on the road.

Tim: We got a ticket for being parked outside our house when they set barriers up around the tree (and our car), which we fought for a long time. It cost them more than the £60 fine and the ticket was eventually rescinded after we legally defended the action.

Rachel: The road closures were so ambiguous. The closure signs went up and down for weeks and we didn't really know where we stood and thought it might not be enforceable.

Jane: Then there was the plane tree at the top of the road near us which I was looking out for. It is outside a business rather than a house and all the premises occupiers wanted to keep the trees, and the property owner gave us written permission to occupy the site should they try to fell the tree. He ended up visiting his insurers who advised that he should go along with the Council decision. He was very apologetic.

Rachel: He ended up signing something with Amey to give them permission to work on the property.

Jane: We looked at the right of way there as a path across the property has been used for decades and goes right by the tree. Also opposite is the supported living property with older people who have that tree for their view. One day one poor man came out in his socks and on crutches and said, 'Thank you, this is my view, I can't get very far. I'll write to the Council'. He had such a strong belief that if he explained to the Council then it would be alright.

Rachel: They said they'd been told that the reason for felling was for access and safety, for the pavement repairs.

Tim: By the time they came for the tree outside our house it was when 'geckoing' [see glossary] had begun as a means of defending trees after the injunction. I stood against our hedge before they set up the safety zone. They asked me to move and I said no. I was encouraged by campaigners that it was OK under the injunction but was worried that others weren't so sure. I bottled it.

Jane: For a while they used to come past each morning. Whether they were baiting us, making us stay out in the cold, I'm not sure. I don't know who was playing the game more, us or them? They may have had no intention of doing anything but driving past to wind us up.

Rachel: We've still got the WhatsApp group for our road from that time. Now if I run out of a cooking ingredient I just ask. We've got a great community spirit among the 'sound' people on the road.

Jane: Even if our binding community spirit is that we really, really hate certain people on the road!

Rachel: This street was an experiment wasn't it? They planted lots of different varieties to see what would work. That's how we've got such a mix. There are lots of random ones.

Tim: I remember back in January there was a guy who I haven't seen since who was inside the barriers and he told me that 'in the tree world this street is quite renowned for all the different varieties of trees, some of which are rare species'. I haven't seen him since but [he] was particularly interested in this street.

Calvin: We had to keep adapting. We worked around the threat of the Trade Union legislation for three months. That was seen off thanks to our defiance and Paul Powlesland's Barrister's Opinion. We then had to get our heads round the undertakings that the council tried to get people to sign before the injunction, followed by the injunction itself. We did bloody well. We weren't just brave and determined, we were clever and inventive too.

Susan Ashworth: When we were 'geckoing' we were being careful not to refer to people's names. Some people did a great job keeping their anonymity for a long time. It was all going well that morning until I went to make drinks and stepped outside and called out, 'Mark, do you take sugar in your tea?', in full earshot of the evidence gatherers and arbs. I just thought, 'Oh, fuck'. We were ordinary people, imperfect, funny, hopeless, defiant and passionate. Sometimes you just had to laugh!

Calvin: Dunkeld Road was the scene of the first open breach of the injunction. I was on a street in another part of the city when I had a phone call from one of the Millhouses campaigners to tell me that two people were inside the safety zone, unmasked, and defying the injunction. The person who called me knew I intended to do the same and I rushed over. I got a lift from one of our stalwarts and when we parked the car the two of us almost ran up Dunkeld Road. On the way I said to my companion, 'Do you fancy coming in too?' and they said, 'Yeah, go on then'. It was a big moment and then there were four of us standing there. It was nerve-wracking and in the films I am bouncing on my heels like a boxer and talking a lot! Council officer Paul Billington turned up quickly. He had given evidence in the injunction court case and was as agitated as me. He filmed me from all angles and we exchanged plenty of opinions. A real point of no return, but it felt good.

Kate Billington: No relation!

Mark: After the Olive Grove protest, Amey had a few weeks when we were somewhat powerless to stop the felling, so talk of breaking the injunction began in earnest. Calvin was keen to do this, and stand up to the Council, and we hoped strength in numbers might be a way to do it; 'there's a limit to how many people the police can arrest', we thought. Unfortunately, we struggled to enthuse more than a few people, so decided just to get on with it one day on Dunkeld Road. Paul Billington came storming up the road to tell Calvin off, but was clearly unsure of what to do about the other three protesters who he didn't know. After taking our photos and a lot of frowning and head shaking he disappeared again, and the crew packed up.

Calvin: A month later, the incident was played out in the High Court. Funny thing was, the safety zone wasn't complete and we technically didn't breach the injunction terms. Only I'd said so much online about what I did, and why, that I still got done.

Jane: The whole thing has really taken its toll on so many people. There are points when I've hit bottom with worry and lack of sleep. Three trees were felled on our road, but several more were saved. Nine months of ups and downs.

Rachel: People put their lives completely on hold.

Tim: For all the people who did things that were well known, there were so many who parked a car under a tree, or checked early morning for crews. All the little things that added up.

2017-2018

'Bonkers'.

Michael Gove, then Secretary of State for the Environment

Kenwood Road is a residential street in Nether Edge, part of the Kenwood Estate laid out in the mid-19th century and lined then, as now, with avenues of trees.

Calvin Payne: On Wednesday 27th September 2017 the protesters on Kenwood Road saw Michael Gove walking down the road, coming from Kenwood Hall Hotel where he had been in a meeting with other protesters, some of whom had been amongst those arrested. He was accompanied by the campaigners and on his way to meet with Sheffield City Council leaders. I'd only just got used to being on the same side as the *Daily Mail*, and I daresay most of the campaigners wouldn't usually expect to find themselves on the same side of the argument as Michael Gove!

Michael Gove [speaking to the Yorkshire Post]: Having listened to people who have been on the frontline, it seems to me clear that the Council has no adequate defence for continuing to cut down trees in the way that it has been. Sheffield is losing, we are losing, an amazingly valuable natural resource and the justification for it seems as flimsy as an autumn leaf. The idea that because tree roots might potentially cause a kerbstone here to be slightly out of alignment, or might theoretically pose a risk to someone's mobility and therefore justifies felling trees that have been here for generations, is bonkers.

Alison Teal: Gove was tremendously well briefed and spoke as knowledgeably about the campaign as any of us could, which blew me out of the water! I was grudgingly impressed by him.

Calvin: We were in a position where a Labour Council was actively trying to imprison some of us and a Tory cabinet member was not only visiting Sheffield, but meeting some of the people who had been arrested during street protests. The Council said they wanted Gove to see the 'reality' of the situation, but there wasn't any tree-felling on the day of his visit. He got the reality and truth of every day on the streets from us, though, if not from them.

Alison: After seeing us, he met with [Council Leader] Julie Dore and [Councillor] Bryan Lodge to try to get them to change policy. He was wildly unsuccessful as the next day they tried to chop down 'his' tree!

Calvin: That was *so* Sheffield City Council! They said that the road closure was in place and the work scheduled, but road closures went on for months sometimes, and the fact that it was that particular tree.

Alison: When they came that morning, there were quite a few of us outside Kenwood Hall on some cobbles which we were confident wasn't part of the pavement or highway and therefore not covered by the injunction. The cobbled area was close enough to the tree for us to prevent the work, so we were inside the work zone but not on the highway.

Paul Selby: On the day after Michael Gove's visit to Sheffield, Amey turned up to fell the tree that Gove had been photographed under, which Council

Officer Paul Billington later told me had been chosen for felling out of spite. It was a Thursday, my non-working day, and we knew Kenwood Road would be targeted somewhere, as it was the road with a closure order for that week. So I was there at 9am, with twenty or so other campaigners, waiting to see which tree they would target. We had a 'garden permission' for the tree from Kenwood Hall Hotel, but we had been told not to publicise it, so as not to drag the hotel into the whole controversy. Interestingly, a segment of the pavement was owned by the hotel. It was demarcated by cobblestones, but to the average passer-by it might have looked as if it was part of the highway. Helen [Kemp] actually sent me the Land Registry record and map for the hotel that proved it. Amey arrived and set up the safety zone and in doing so they encircled the campaigners standing on the cobbled part of pavement and continued the barriers onto the hotel land. At this point, some campaigners nervously stepped out of the safety zone, fearing they would be found guilty of breaching the injunction. Staying on the Kenwood Hall-owned land, I called over the lead arborist and quietly told him that his crews and his barriers were trespassing on Kenwood Hall property. He laughed out loud and prepared to step away to continue preparing for felling. Then I showed the lead arborist the Land Registry map on my phone and he started to look nervous and went to consult his colleagues. After ten minutes of discussions, they moved the barriers out of Kenwood Hall-owned property. The campaigners were now confident, and all the rest of the protesters who had moved away after being threatened with the injunction moved back onto the Kenwood Hall-owned land. The sheer number of campaigners in place under the canopy of that tree meant it was impossible to even take the roadside branches safely.

Calvin: I was inside the same zone but on the road. I was committed to doing as many things as possible that might end up with me in court. I was determined to force the issue, but had also slightly lost the plot by that stage.

Alison: A group of us had a conversation and we knew that we were within our rights and not on the highway. We asked the hotel for confirmation, but anyway it was clearly marked out as separate from the pavement. I think the Council thought they could get away with anything by then and that the injunction gave them confidence and they overplayed it. It was hubris.

***

Alison: The next time the crews came to Kenwood Road it was to the other end of the road, and the first thing I saw was that the work zone was enormous! There were lots of us strolling around inside an incomplete zone. They didn't have any barriers alongside one of the walls, and we knew the injunction made it clear that private property couldn't form part of the work zone.

Helen McIlroy: We saw that they had left a gap between the Heras barriers and a garden wall. Chris [Rust] and I got into the space. I sort of wriggled in, although I may have pushed the barriers out a bit too. Darren Butt [Amey account manager] said to us, 'You can't do that' and took photos of us. I didn't want to fall foul of the injunction so I asked him, 'If I come out from here will you delete the photos?', and he said he would. I later saw that it formed part of the file of photos in the injunction case, so it looks like he didn't do so.

Calvin: I was inside that zone in a slight daze too, although there is a video on YouTube where I seem lucid enough, and spoke well when asked what was happening.

Alison: I was inside the zone that day and I was confident I wasn't breaching the injunction. I wasn't worried. I was there for quite a while until Darren Butt came up to me and said, 'You know you're in trouble, don't you, Alison?'. I said, 'What do you mean?', and he said, 'Well, you're in the work zone and breaching the injunction.' He was essentially trying to intimidate me. I said, 'No, it's OK, the barriers aren't complete'. I showed him where and why. He just told me, 'Oh, that doesn't matter, we've got you now'. So I left the zone, saying 'OK, but I don't think you have.' I left then, as there were others inside the zone and the tree was protected, for now at least. Butt was pleased with himself, he was smiling.

Helen: I filmed Alison leaving the work zone when she was asked, but it wasn't needed in court as we now know.

Calvin: We were both charged with contempt of court for breaching the injunction over these events on Kenwood Road. Our cases were different, as were the outcomes, but the events on the road were to be examined legally just a few weeks later. They must have thought that they had Alison, they certainly seemed confident, and they were gutted in court when the charges were thrown out.

Alison: The two incidents that I was confident about ended up in the High Court. The first one, the Kenwood Hall cobbles, was thrown out very quickly. The Council legal team admitted that they couldn't prove that the land was part of the highway. Then evidence was presented by their barrister accusing me of being in a work zone in breach of the injunction. Darren Butt and [arborist] Jason Wignell both said in their court statements that I was in the zone at a time I wasn't even present on the road, but I had photos which proved that I wasn't there. The judge, Justice Males, then explained to the Council legal team in court the same thing I had tried to tell Darren Butt at the time. He told them how the injunction defined a work zone. Something we already knew.

Calvin: Both of the occasions that I appeared in court over, on Kenwood Road and Dunkeld Road, there were several people inside the zone doing the same thing as me. I'm glad they didn't get any bother, but it does show that the Council and Amey had their targets.

Alison: There were any number of people they could have taken to court from what happened that day, but they picked me out. It was political and all the more shocking that it involved a private contractor singling me out too.

Calvin: That day in court was a strange experience. I don't remember large parts of it. There were three of us at the start, then Siobhan O'Malley's case was dropped early on. Alison's case then took up the rest of the morning with the Council and Amey witnesses presenting their evidence. I was so pleased when Alison's case was rightly dropped. We went out for lunch with supporters, talked to the media, then it suddenly occurred to me that I was on my own in the afternoon!

Alison: The whole thing about me 'getting off on a technicality' was disgraceful. The Council decided at officer level to spin it that way. Far from being 'a technicality' I observed that I was in the right on the day and acted accordingly. I'd pointed out to Darren Butt that the zone wasn't complete. They, and we, could have saved a lot of money and trouble if they had listened.

Calvin: The strange thing about Kenwood Road and the possible consequences for us in court was that so little really happened on the day in question. When you think about the scenes that followed on Abbeydale Park Rise and Meersbrook Park Road, it was all a fuss about not much.

Alison: It was all very uneventful really, wasn't it? There was no chasing, swearing, climbing, none of the things that were to follow. My sense in court was that the incidents involving me were so incredibly low-key. It was hard to grasp how something so minor became such a big thing. They made it so by taking it to the High Court. It seemed like nothing at the time, and in court it turned out to be nothing!

***

Sheldon Hall: Residents organised night patrols, keeping watch on the streets from the early hours until dawn, when the regular rotas took over. I volunteered to patrol on foot (I don't drive). So for several weeks in December and January, no matter what the weather, I found myself walking the streets of Nether Edge in the early hours. I really rather enjoyed those walks. The stillness and silence of the night meant that it was possible to hear for miles around. Cats and foxes would stick their heads up to wonder what I was doing prowling their territory. Often I'd see other campaigners patrolling in cars or on bikes, criss-crossing the area and stopping to chat or wave. On one occasion I saw one sauntering down the road, with an extending ladder over his shoulder and carrying a bag of bat-boxes to attach to trees. Only one night, on Kenwood Road, did I see any serious action. There were four or five arbs; one of them was up the tree while the rest stood around collecting the cut branches. They were working only by the light of a nearby street lamp and the torches on their helmets, with a tape cordon around the work area – in other words, not a proper safety barrier. They saw me coming, I put out an alert and within minutes two other campaigners arrived on the scene. I had been circling the cordon trying to find a point of access but was shadowed by the standing arbs, blocking my entry. David [Glass] brooked no resistance, stepping over the tape and into a neighbouring driveway under the tree. By the time Chris [Rust] appeared it was all but over, and the arbs were packing up.

John Baxter: I got back to Kenwood Road one evening that week after the arbs final deed was done. There was a lot of 'slow walk' protesting as they were moving out equipment. My stand-out abiding memory of that day was of Jim Steinke, the Labour Councillor for Nether Edge, just standing there, just watching. He was there a long while, frowning, not interacting, watching us disapprovingly. I remember scowling at him, he was our local county councillor in Netherthorpe decades previously. I wasn't a fan then either.

Klive Humberstone: On 16th December 2017 I had just played a sold-out show at the O2 Academy in Sheffield with the Everly Pregnant Brothers. Now four days remained before Christmas and we were all wondering when Sheffield City Council, Amey and Acorn would cease activities for the festive season. News on the campaign WhatsApp groups reported that barriers had gone up on Kenwood Road at 5:55 that morning and arborist crews had taken the opportunity to hand-cut branches from the canopy of trees in an early-morning raid. I grabbed a snack bar, my coat and hat, and made my way to Nether Edge as soon as I could, walking there with my dog, Harvey.

Simon Crump: We got to Kenwood Road early, at half past five in the morning and discovered that the bastards had been in the night with hand saws and cut all the branches off a line of lovely trees at one end of the road. The branches had been left on the pavement surrounded by an incomplete ring of plastic barriers. To me, this seemed like a new low. My first instinct was for us to drag all the branches onto the road and then call the police, blaming the arbs for the obstruction. I started dragging the branches into the road and a few other people joined in. As soon as we'd finished I was persuaded by more reasonable, responsible adults that calling the police was a bad idea, and that instead Amey having to come back and remove the branches would at least tie up a crew and a chipper.

Roger Doonan: It was ridiculous, branches left there, no cones, no hazard lights, nothing. We were being very nice, very 'tree campaign', discussing what to do. It was a blatant hazard.

Jenny Saul: When the call went out at 6am to let us know that they had been cutting branches in the middle of the night I headed out to Kenwood Road. Eventually I went home to get my then twelve-year-old son, Theo, and take him to holiday camp. My plan was to walk by with him, drop off some baked goods for the protesters, then return to the protests.

Simon: A crew turned up soon after that, presumably to fell what was left of the trees, and at this point, I think more because we were pissed off with them than anything else, we decided to make a nuisance of ourselves, to make their job as difficult as we could, to slow them down. So a number of us got inside the barriers and some others got themselves on the pavement between the barriers and a perimeter wall. The arbs began chipping the brash at a distance from us. They were joined by other crews, and eventually by the police.

Jenny: When Theo saw what was happening, he didn't want to leave. He asked if he could go inside the barriers to bring baked goods to Simon, who was sitting on the branches. Everything seemed very safe, so I let him. The arbs and security were asking Simon to leave, but only asking, and everything had been kind of in stasis for hours. It was unthinkable that they would do anything violent with a child in there. So Theo went in, he recalls it was really fun sitting in the branches sharing treats with Simon and talking about guitars. But then suddenly things turned chaotic as they started pulling the branches into the chipper with Simon on them. Theo remembers feeling scared and angry. I was totally shocked and horrified that they would do something like this. When the video went viral, Theo says his initial reaction was that this was really surprising but it was 'cool to spread the concept of branch surfing nationally'.

Simon: As they got closer and closer to us, there was Jenny Saul's son Theo, an older chap whose name I never discovered, Klive with his dog and me, all inside the barriers. I'd spent a nice hour or so with Theo beforehand, talking to him about guitars and rehearsing him to say the line, 'You'll never take me alive copper!', when the police arrived. It was also the first time I'd ever met Benoit [Compin]; he performed a poem which, I think it is fair to say, nobody understood. Then it really did kick off. We were basically now the only obstacle standing between Amey and them getting the job finished. The arbs (Jason Wignell was one) started pulling the branches from around us and feeding them into the chipper. We were hanging onto the branches and I don't know why I did it really, but I threw myself onto a big bit of branch and clung to it as they tried to feed it towards the chipper. I could feel the heat from it and the noise of the engine getting louder and louder as I got closer and closer to it. I thought that it was brinkmanship on their part, at least I sincerely hoped it was, and at the crucial moment before they fed me into the chipper, they overturned the branch and flipped me over onto the road with the branch on top of me. Which was a relief. Having said that, they could easily have damaged my spine by doing that and I think that they'd lost their tempers at that point. I was trapped and it took several fellow campaigners to lift it off me. Benoit was first to come to my aid and I distinctly remember shouting over the din of the chipper, 'No! I don't wanna be rescued by a fucking poet!'.

Roger: I came back a little later when Simon was fighting amongst the 'twiggage'. Just how far off the normal scale of events was all that? The night raid forced us to respond. I thought it was brilliant really!

Benoit Compin: I did see that a few people were trying to raise attention on the matter by doing some funny things, and I thought, 'These guys are daring, that's cool, but they may not have my experience of entertainment and misdirection on the ground. They can probably do with a little help because they were struggling'. Then I thought, 'I have something to add to this', and attended the protest on Kenwood Road, wasn't it? The chipper thing.

Calvin: That was the first time I remember encountering Benoit. Suddenly there was a French guy reading a poem and I thought, 'What's going on here?'. I joked about it when I spoke outside their court case. Simon was being Simon and I didn't know Benoit, who was really quiet compared to the Benoit we came to know later!

Benoit: I was testing! I saw what Simon was trying to do, creating stories as well. And when I saw that I decided I needed to go and see what I could add.

Roger: People seemed to be taking the line that we always had to be better than the other side, but I was intending to call the police and report the obstruction the arbs had just left lying around. That morning was just one example of how mental it all got. Just mad.

Simon: Russell [Johnson] filmed the whole thing on his phone and put the video on the STAG Facebook page that evening. The best thing about the video is that at one point Calvin can clearly be heard shouting, 'Oi you prick!' at Jason Wignell. The *Daily Mail* got hold of it the next day and suddenly it had 3.8 million views. It went viral as the young people say. I think what some folks don't fully understand about this form of NVDA, which was described as a 'stunt' by several of the 'keyboard warriors' on the STAG Facebook page, is that whilst it is exactly that, it is a 'stunt' which delayed four felling crews for most of the day, when they could have been elsewhere attempting to fell trees across the city.

Calvin: Part of me was thinking that if Simon was 'chipped' then that would be the end of the felling programme there and then. I told him afterwards that it would have been 'a short term loss for him, but a long-term gain for the campaign', but he didn't see the bigger picture!

Klive: A crowd of campaigners gathered along the pavement. Acorn crews were present with chipper vans, along with a couple of Amey 'evidence gatherers' filming proceedings. Barriers had gone up around the branches and the brash that had been sawn from the trees earlier that morning. I think three or four chippers were parked up along Kenwood Road. It seemed that they were not going to attempt to fell any of the trees, since the number of campaigners was fifteen or twenty strong. It became apparent that the crews wanted to 'chip' the cut branches before moving on – we simply tried to 'stall' them from doing this so that they couldn't move onto another location in the city. As the arbs attempted to remove and feed the branches into the chipper, a number of campaigners stepped into the work zone and held onto the branches. At about 10.30am, a police constable turned up, simply to 'observe'. I was holding onto a branch, with my dog beside me, inside the work zone. I distinctly remember the constable scowling at me and shouting, 'Stop that, or I will report you for animal cruelty' – the situation seemed quite bizarre! But we knew that by delaying the crews and trucks for an hour or so, we would hopefully stop them from felling another tree elsewhere that day. I let go of my branch and moved out of the work zone onto the pavement. Simon however remained inside the zone and held on defiantly to a large tree branch. I then saw one of the arbs pay particular attention to the branch that Simon clung to, trying his best to drag it towards the chipper before dropping the heavy weight to the ground. Simon was twisted to one side and became entangled in the branches whilst still holding on tightly. Voices were raised, photographs were taken, footage was filmed, the police did nothing. The arb crews packed up and left; no further fellings were made that day.

Simon: I should add that I'd been reading quite a bit about situationism and was particularly interested in the situationist concept of detournement, which in my world translated into taking hold of a situation like this, escalating it and then pushing it to its logical extreme which is exactly what happened here. Or as Calvin later put it, 'Or maybe you were just being a bit of a twat'.

2017

'I've never seen anything like it. I don't think I'll ever see anything like it again.'

Mark James, Resident

Edgedale Road is a tree-lined residential street in the Carterknowle area of the city, and is adjacent to an air-pollution blackspot.

Sam Cathan: I was aware of the Tree Replacement Work notices on the street, but I also had to endure what was done on Swaledale Road [a neighbouring street with predominantly pro-felling residents]. That lit the fire for me, and I could absolutely see what was coming. I was devastated by what played out there and was thinking, 'No! We've got to do something about this'. The people here were absolutely ready and not going to take it lying down!

Mark James: I'd got into some interesting conversations on Swaledale Road. One of the people turned out to be an Amey worker, because I spied him later on Meersbrook Park Road. I've never seen anyone's colour drain from their face like his, because he was in his civvies when I saw him on

Swaledale and he was swearing and shouting and everything. I can only assume that he lived on Swaledale Road. He was shouting at me, 'We want the trees down' and then a few months later, I caught him. I just went, 'Oh, so you weren't an innocent passer-by, you knew exactly what you were doing'. He went as white as a sheet. Edgedale Road has nine trees and they were gonna take all of them, bar one, so there'd be only one tree left on the road. When the tree panel re-surveyed it, they said to save all of them, bar this one right outside my house, because of the pavement.

Sam: I remember one of my neighbours being in favour of the felling. I didn't agree with his position but was willing to understand where they're coming from, but the thing that pissed me off was that he started talking about the leaves in the gutters, and then he went on to talk about the trees causing moss. This is what started riling me, he was making this connection between the trees causing moss on his roof and I said I don't think that's the case. I wasn't being aggressive but I wasn't polite either, and remember telling him that I didn't think he knew what he was talking about. I suggested he read a few more books and then we'd have a conversation. I regret it and I was being arrogant, being passionate and annoyed, and I would like to apologise to him at some point because it's his point of view and I could have been a lot more diplomatic. It has made relations slightly frosty.

Mark: One of the things that struck me was on the first day that they were due to carry out the fellings I took my dog for a walk early in the morning and there were people already out on the street waiting. The road-closure signs were up, there was no sign of Amey yet, and yet there were people out on the street. It was quite clear that, contrary to some of the stuff that was spoken on Swaledale Road, that the vast majority of people on this street understood the dynamics of the street and what the trees were doing for our benefit, rather than to our detriment.

Sam: We were twitchy, very twitchy.

Mark: On the first day of the week-long road closure nobody came, same again the next day, not until Wednesday did the crews arrive. And when they did come, 'it was just like Vrroom! Out and up!'. We were ready.

Sam: I live on the other side of the road to Mark and we spent the entire summer listening to those trees on Swaledale Road getting chopped down, to the point where I couldn't, and it sounds really dramatic, but I couldn't sit in my garden. I spoke to my neighbour Jan as well, we felt hopeless just waiting. I bumped into a couple of people from this road and we went to stand on Swaledale and support the people there, and what we were greeted with was something very toxic. It was very uncomfortable. That's when I got the taste that this was going to be difficult.

Mark: They came for the tree at the top, the one outside Sue's house. We stood on the wall, and when they tried to hand her the injunction breach threat letter, she refused to take it. The first contempt-of-court case for alleged injunction breaches was about to happen, so they had evidence gatherers there. It became really clear that it wasn't going to happen – they were not gonna get these trees. I've never seen anything like it, I don't think I'll ever see anything like it again. There were people defending the streets, their streets.

***

Calvin Payne: I remember at the bottom of the road when they tried to serve injunction papers on some of us for standing or sitting on the church wall (which wasn't covered by the injunction, because it wasn't part of the highway). So we were giving them a bit of stick for that. It was a few days before my court appearance in October 2017, and one of the Amey arbs came over to talk quietly to me. I couldn't go anywhere as I was perched on the wall. To my slight surprise he wished me well in court and said that no-one in his team wanted anyone sent to prison, which was a very real prospect at the time.

Sam: They were working on the tree by the church wall and the tree had a bit of a trim on the side facing the road while a few of us were perched along the wall. They tried to give me a piece of paper but I wouldn't take it. No part of me was on the highway, and I knew enough about the injunction to be confident. The evidence gatherers said I was in breach. I replied, 'By all means, if you feel the need, you can read it out to me'. I tried to use humour to deflect the situation, and commented on his reading style and gave him a few tips.

Calvin: One of many such surreal moments after the injunction. We've argued with the other side that parks, driveways, church walls, and more, aren't part of the public highway! The injunction was clear on that.

Sam: When I stood on the wall with Calvin and the others it was going through my head, 'I work for Sheffield City Council'. Here I was thinking, 'How far am I going to take this? Am I prepared to put my job on the line?'. At that moment I was suddenly realising how high the stakes were. When I stood there I also pretty much decided to come off antidepressants after twelve years and I've stayed off them since. There was a lot going on in my head while standing on that wall.

Calvin: It's very empowering to do something that makes a difference. Any gap you fill in your life and anything you want to prove to yourself, that's great. Lots of us have. It's been positive for so many people, including me, for more reasons than the very good one in front of us. We tested ourselves and others.

Sam: People on our road did amazing things that week. When 'K' had gone in the barriers and was one of the 'bunnies' [see glossary], I was just in awe at what she had done. It was incredible. The fact that she had a young baby and her husband was holding their tiny baby while she was sitting there in the cold refusing to let the tree be felled. Whilst I was watching this going on, the neighbour who I had clashed with started having a go at her, talking about sap on his car, which I was furious about when I compared what he was saying to what this brave woman was doing.

Calvin: I'm not sure if it was the same person, but I remember one day when a woman I'd never met before came up and whispered in my ear, 'I think I can get under those barriers'. She was slightly built but I still didn't think she could. As I was thinking that, she was straight under the barriers and inside the work zone!

Sam: So we had residents with differing views, the arbs and the evidence-gathering people, all out on the street together. A lovely woman neighbour, very unassuming, very quietly spoken, had a teapot in this lovely knitted tea cosy. The evidence gatherer told her that she was breaching the part of the injunction about 'encouraging' protest actions by making tea for protesters, and threatened to take a photo of the teapot!

Mark: One of the things was that, when those evidence gatherers were going up the street and I was standing with Milly, my dog trying to calm her down, is that my wife would have been in hospital at the time. She might have been downstairs, and they were filming us through the window! This is utterly outrageous, in our own house!

Sam: I couldn't go into my house because I recognised that one of the evidence gatherers was sitting in his car waiting, watching me, waiting to see which house I was going to. So I ended up wandering around the block. That had never happened to me before, and as a woman on my own, it was very unnerving.

Mark: I mean, I was a Labour Party member at the time, it was my own party that had instructed people to film us on our own street!

Calvin: It was the first time that I remember a concerted action happening on a road where a (then) Labour councillor lived. That brought things into sharp focus for residents.

Mark: He is a decent bloke on other issues, but when it came to discussing the trees, they [the Councillors] would resort to tribalism. They thought they were dampening down opposition when, in fact, they were fuelling it.

Sam: I gave the councillor the opportunity to give me his own personal feelings about it because we both have kids in the same school and I remember saying to him, 'We both have children in the same year'. I basically asked him how he feels about his daughter growing up on this road without the trees, and I thought that'll be a good way in. He just regurgitated the same old party line. He just couldn't do it.

Mark: Probably because it's my field of study, I was always noticing the way class, race and gender dynamics played out in Technicolor on the streets. One of the things that played out on Meersbrook Park Road, in the late Autumn when Amey first went there, was that there were a couple of working-class women who shouted at one of the arbs up the tree, 'Are you getting a hard on doing that?', and it worked. Because he stopped. And one of the things that struck me about that, was the portrayal of the protesters being 'polite and middle-class'. I felt middle-class women would be less likely to say that out loud, they might think it, but would be less likely to say it. But these working-class women, young working Sheffield women, they used every device they had, and it went like a knife through butter and it was so funny because it was just a little indication that the class dynamics of the protests were much more nuanced than the Council were portraying.

Calvin: The way the Council saw the issue of the protests in class terms was interesting. We started off as middle-class and self-interested but then, when it suited them, we became a bunch of layabouts and hooligans instead.

Mark: People who did not see themselves as protesters, in terms of confronting the forces of the state, of the law, changed on this street. Quiet, unassuming, well-educated, middle-class people turned into committed activists. And it happened right there – outside my house!

Calvin: Every attempted felling, every street, has seen stories played out. Hundreds, thousands, of stories of people doing things that not only had they never done before, but they'd never even thought they would ever do.

Mark: What happened on this street was just incredible.

Sam: I've always assumed that what happened here was a replica of what happened on other streets.

Calvin: So each time, on four visits to this road, the fact is that they've been prevented from doing what they came to do by the actions of people, and often by the actions of *one* person. There have been two instances of what people have done on this road, and no one else knows who they are. Two people on this road, two *women* on this road, have masked up and one's gone over and one's gone under, but they both got inside and stopped the felling. And that's incredible, really, when you think about it, we've seen it enough times but on its own, if you take that into isolation, it's an amazing act. Happening once is a story, happening hundreds of times over a two-year period across the city is remarkable, sustained, peaceful resistance.

Mark: I think that despite some differences, the atmosphere on the street between neighbours is better now than before. People have interacted.

Sam: I know my neighbours better now, and have met so many people through the tree campaign, but when we had a campaign social, my husband asked me, 'Are you going to go?', and I was like, 'My arse I am – I don't want to spend an evening as well as all day with them'. I have my limits!

2017-2018

'And so it went on, "Cunt, cunt, cunt. Cunt", and all the while him smashing the barrier against me.'

Simon Crump, Protester

Lismore Road is a short, steep, tree-lined street in the Norton Lees area of the city.

Maggie Young: We've lived on Lismore Road for over twenty years, and there's a picture of our house with the tree that we tried to protect outside it which was taken in 1905, shortly after the house was built. We vaguely knew there was going to be a consultation, in fact, we eventually found the plain brown enveloped council survey and it was quite complicated, but we did go onto the computer and talk to a few neighbours about it.

Martin Young: The language was a bit like those A4-sized things that they staple to trees, those laminated things, that are a bit vague, really vague, yeah, 'in the next two weeks...', whatever, and all that crap. They say it's highway maintenance and make it sound like they might be doing your pavements, without absolutely telling you that they're going to chop down your tree.

Maggie: Basically we got the idea that there were eighteen mature trees on this street and that they wanted to take nine. So we talked to a few people on the road and the consensus of people I spoke to was that nine was far too many. Maybe take two or three, but nine is too many.

Martin: I remember that we did the survey online and basically said that we didn't agree with any of it. We wrote quite a nice, eloquent reply to it and we simply gave our reasons why we didn't agree with the fellings. We didn't say that we're against tree-felling completely, we just thought that the reasons given didn't justify the actions. There were a few of us on the street who were opposed, probably five of us, actively, *really* against it on this road, but most other people were either sitting on the fence or actually in favour of it. It's an interesting thing, because when all this starts, it creates divisions between people. It's like suddenly people over the road are on the opposing side. A neighbour of ours that we hadn't had any problems with before was rude to us on so many occasions that I just don't have any time for him anymore.

Maggie: Our neighbour above us, he definitely wanted the tree outside his house gone. He'd been asking the Council to take his tree down for fifteen years and he kept saying to us, 'You're not going to stop me having my tree down, are you?', and I said, 'Well we won't stop you having your tree down as long as you don't stop us having our tree up'. He knew what we meant.

Martin: He wanted it down so he could get on his drive but the thing is, he's not parked his car on his drive once since it's come down.

Maggie: There was a woman from up the road, who suddenly appeared when they were taking a tree down, and I said, 'What do you think about all this?', and she moaned about aphids coming in her window. She said, 'Listen, my son is thirty-two years old and in that time, the Council has never been here. They've not pruned one single branch in all of those years. If they'd done something sooner, it would have never come to this'. We feel that the tree and the house, they've been together, partners, for over a hundred years. Also that means if the tree is moved, well who knows what happens to foundations and walls?

Martin: I asked one of the Amey managers that question when he was outside our house on the second to last time they came to our road. I asked him that, and Jaq [Pixelwitch] was filming it for her documentary, and he just wouldn't answer the question. I asked him, 'Is that a risk you're prepared to take? If you cut that tree down and the roots rot, it might move the foundations of the house and cause serious damage'. They didn't factor in the risk at all.

Heather Russell: One morning, I was there shortly after nine o'clock and there were six arbs and barrier people, and also one person who was there to take photographs, an 'evidence gatherer'. I went up to the barriers and we did the 'barrier dance'. We went backwards and forwards and round and round, and I finally got around this guy who was trying to stop me getting inside the work zone. At that point, he actually tried to get the other members of the felling crew to crowd around me, and they all refused. They all hung back, and then he started saying, 'You are being filmed', as he put this little camera up on the car that was actually inside the barriers. Then he came up to me and stood right in front of me and my face was level with his chest, and he said, 'You are invading my personal space. I feel intimidated', to which I burst out laughing, and then he got really mad! He picked up a plastic barrier, rammed it into me, sort of whacked it into me, and then rammed it down on my feet, just to get the edge bit right down on my foot, at which point I started shouting 'Assault!'. I was there by myself and I wasn't going to stay there. The arbs still hung back and did not look happy. I went off and I reported it to the police. When I reported it the police even said basically, 'Oh, it's you', and they said, 'Well, what do you want this time? Should this go to court, should it be a caution or what?'. I said, 'Actually what I really want, I want the whole thing to stop. I don't want this man to be assaulting people'. The following day they were on Lismore Road again and when I went there, I didn't say anything to the arbs to begin with. I just sort of walked around. I could see that I could get into the work zone by climbing up a wall and then hopping over this little fence. Which is what I did, and one of the arbs came in and said to me, 'We don't want you doing that again. If you need to get in through the barriers, I'll open them up!', and then he apologised for what had happened the previous day. He said that they were very uncomfortable with what had happened to me and that they were sorry. I understand that the guy I'd reported didn't come back onto Lismore Road, and wasn't seen at all after that. So I assume that he was spoken to, moved, or even lost his job.

Simon Crump: Sometimes that did happen in those more easy-going pre-injunction days. Familiar faces on their side would mysteriously disappear from the scene after losing their tempers with peaceful and good-natured protesters.

Heather: I arrived on Lismore Road another time and they'd already chopped quite a few branches, and I got in with all the felled branches and it was quite funny because I was literally buried up to my waist, and the guy was fantastic, he came sort of wading through the branches and said, 'Would you like to move over there?'. I think it's the 'old lady with white hair' thing that happens, and you know I'll happily play on that to stop things happening, and also to make them behave reasonably.

Calvin Payne: On our side of the barriers, the protesters were sometimes a little older and often female. Their staff were almost entirely male, and fit guys. A lot of the altercations between them must have looked ridiculous to people watching, completely one-sided. The introduction of the SIA's, nightclub bouncers who were used to dealing with drunks and people being abusive, highlighted the difference even more.

Heather: On another occasion I'd stayed under a tree next to a wall for seven and a half hours. I really needed the loo, and I stood there, crossing my legs! That day, Maggie [Young] came up with a cup of tea, and the two security guys were on the far side of the barrier, inside it, and I was obviously outside the barrier against the fence, and I leaned over to get the cup of tea and that's when they came rushing at me and forced the fence in at me. I put my hand up and actually sort of went, quite dramatically and deliberately dropped my phone, and they were saying, 'You're assaulting us', because I'd put my hand on his chest on the other side of the barrier. They were going to call the police and I just said, 'Oh, I've got to get my phone!', and sort of dived at the phone and they stopped and were fine after that. I never did get that cup of tea!

Maggie: I was going away to Portugal for a week and that sodding week was the exact one they chose to come here! By now I was quite frantic, because we'd given the tree at the bottom of the road a name. We'd decided to call her Linda. And once she had a name, that was it, we definitely wanted to save that tree! Enough was enough. I decided to make a rota of all my friends to be there the whole time I was on that week's holiday; neither of us were going to be there for the whole of that week. Some of them were people we knew, but some were people I'd never even met before. People just volunteered to come up and be in this house, and defend the trees for the whole week. I had three or four keys cut, I had a person for every day, and I kept texting and emailing them all from Lisbon and saying, 'Did you go?' and 'What happened?'. We had to get through five days, and when I got back I said, 'That's it, I'm not going away again'.

Martin: When we were both away, we had all these people on rota coming round and they were fantastic, they left little A4 sheets, reports of what had happened. Every day I'd come back expecting to see a tree gone. Before the injunction it was great. You felt so powerful didn't you? So easy to just put a toe inside the barriers to stop it, the golden days. The first time felling crews came to Lismore I was meant to be working. And Simon [Crump] turned up in a taxi and they'd come and set up. They only had one set of plastic barriers, and they'd set up across the street from us, and we got some kitchen chairs and put them inside the barriers and I played guitar. They just hung around watching and listening. They even shouted requests. 'Wonderwall' was one! That was the first time I did that though, we just jumped inside the plastic barriers and stopped them. The next time they came, it was horrible. It was raining as usual, and they'd set up barriers around a tree further up the road. But they were those big metal Heras barriers, it was after the injunction, the middle of October. I went up there with my camera and said, 'What are you doing? There's no road closure signs, do you have permission?'. They said that they did, and that I should clear off. It's a bit different when you're by yourself. They bully you, don't they? They said that I didn't know what I was talking about and that I was an idiot and to leave them alone. I kept on asking all these annoying questions. They said, 'If you don't like it ring Amey', so I got on the phone and called the office, but it took ages to get through to anyone. It was raining and I had water damage to my phone and my phone turned itself off. I was hanging around without getting inside the barriers, I was trying to get as close to the tree as possible. The arb up the tree shouted down, 'Can you fill in a Health and Safety report, or something, from "Alison" Young?'. He was just trying to be all smart, like, 'I know all your names', they did a lot of that.

Maggie: Then there was the drama with my car. I had quite a beaten-up, battered old car, and the day before they were due to be coming again, the day before, my car just naturally all by itself got a flat tyre, and I thought, 'Oh well, good!' I won't bother breaking an arm and a leg to get that mended. It can just sit in the road outside the house with a flat tyre and if someone says I've got to move it, I will say to them, 'I can't, it's got a flat tyre.' That was a Monday and that was when all hell really broke loose!

Simon: That week was when the SIA goons were employed for the first time. I remember seeing them walking up the road. We'd never seen anything like it at the protests before then.

Martin: That was an absolutely bizarre sight, them arriving on the road. They came in sort of all guns blazing and thought they could use 'reasonable force'. It was a horrible, horrible, day and they were really over enthusiastic, the SIA guys, weren't they? Then I remember I was standing at the tree up from Simon and they were shoving the fence up against him and he was trapped between the garden wall and the Heras fence.

Simon: I got the callout on WhatsApp that they were setting up on Lismore and I went over. It was the tree opposite Martin and Maggie's at the top of the road, and it was the big metal Heras barriers. I got in with a few seconds to spare, between a low garden wall and a barrier as close to the tree as I could. I had to run along the top of the wall for a bit to do this and then wedge myself in, it was a bit of a struggle but I managed it. Instantly, two burly Scottish SIA's got inside the barriers opposite me and just leaned on them from the inside, crushing me, which was bad. It soon got even worse when they hit on the wizard wheeze of tilting the barrier back and smashing it into me over and over again, and with each impact the bigger of the two (who I later learned was the head of the SIA company) was muttering 'cunt' every time they hit me with the barrier. And so it went on, 'Cunt, cunt, cunt. Cunt', and all the while them smashing the barrier against me. Eventually they hit me so hard that my hip bone took one of the coping stones off the top of the wall. At this point, the resident whose wall it was and wanted the tree removed, came out and joined in the fun. I've been in a few situations like this and generally, in a strange kind of way, those situations can be quite entertaining at times. This was fucking definitely not one of those times. Eventually, they seemed to realise that a repeated assault on camera wasn't going to work today and completely lost interest. At which point Martin joined me with no problems whatsoever. He had his guitar with him, as he did under the same tree during the summer in a less violent time, and we had a bit of a singsong. Talk about one fucking extreme to the other.

Calvin: I'd seen on Facebook that Simon was getting beaten up, so I went over to watch. This was far too good to miss. By the time I got there, quite a large group of protesters had gathered on the opposite side of the road and to my disappointment, Simon was no longer being beaten up. He was singing songs with Martin. Shortly after that a bunny went over the barriers to huge applause, which quickly turned to howls of protest.

Simon: As soon as the bunny was in, the SIAs were on him. It was horrible, really brutal. He was on the ground face down, trying to protect his head with his hands, and they were just roughing him up. I cut the cable ties on the barriers in front of me and went in. I don't know why really, just the heat of the moment, there would have been nothing I could have done. On the way back out, I accidentally caught one of the barrier guys on the side of his head with the corner of the metal barrier as he tried to stop me from leaving the work zone, which was a shame. Jaq's got footage of me entering the work zone on that day, she didn't put it on YouTube, but I was still expecting a knock on the door from the process server for several weeks after that. The fact that it never happened, I suppose, is that to get me done for it, they would have to have shown the court a film of what those SIA's did to that bunny.

Calvin: Everything was happening very quickly when Darren Butt ran past me towards the work zone. 'Panicking, Mr Butt?', I said, and he replied as he ran past, 'I never panic, Calvin', but he was.

Martin: It was awful, because they messed Simon about, and then I got in with him, because two is better than one. We were playing the guitars and all that kind of stuff and while I was up there, next thing I knew is that a tow truck appeared and I looked, I couldn't quite see it, and I just saw the sort of arm of the tow truck. I ran out, and the next thing I knew was they got the car onto the ramp of the truck, and that's when I just lost it at that point. I think Jaq's got that on film! I ran inside to get the car keys, came back and stood in front of the tow truck and said, 'I'm not moving until you take the car off the ramp' and the tow truck driver said, 'I can't take it off, if you want it to go off, you speak to the parking officers'. The parking officers were sitting in their car inside the closed-off zone. They'd fenced off their car anyway, and then the police came because I was blocking the tow truck. It was me that rang the police then, because the parking officers wouldn't speak to me. We were knocking on their truck window. They were literally just ignoring us. They were looking straight ahead and they wouldn't speak to us.

Simon: My last memory of that first day on Lismore Road was of the boss of the SIA company lying to a copper about me. I'd cut through the cable ties with a penknife I've carried since 1975 (I use it for getting horses out of boy scouts hooves) and after that one of the SIA's was making a bit of a deal about me carrying a knife, so I handed it over to the copper to avoid any complications. About an hour after that, the boss of the SIA company approached the copper, and it really was a BAFTA-winning performance

'He's just threatened me with a knife!', he said. 'When was this?', the copper asked. 'Just now', the SIA boss said. 'Was it this knife?', the copper said, showing him the knife I'd handed over.' 'Yes, that one! He said he was going to stab me!'. 'Simon gave me this knife an hour ago', the copper said, and turned his back on him. 'I'll have you Simon, you cunt,' the SIA said, and then he fucked off. So that was a nice day out.

2017-2018

'While stuck there geckoing, I started breastfeeding. I had my dog with me, and I had my baby with me. It was all very nice, quiet, and calm.'

Elizabeth Mountain, Protester

Spring Hill Road is a tree-lined avenue in the Crookesmoor area of the city.

Kaarina Hollo: The campaign story goes a long way back on Spring Hill Road. It's a beautiful road. There are seventeen lovely trees on it and they wanted to fell eleven of them but only managed two, one of which was diseased. It's always going to be tied up with Dronfield Tree Services [a local contractor working for Amey] for me because of the tree on the corner with Bower Road. The guy who lives next door to it, Rod, told me it was probably sixteen or seventeen attempts they had made on that tree.

Melia: The one outside the house of the resident that gives us the parking permits? That tree, that's what got me interested in the issue and the campaign. I used to walk home from work along Spring Hill Road.

Calvin Payne: I'd also been there a couple of times, both early mornings; it must have been when they had the road closure. I used to walk along the road from home to work too. It's a short but very nice road with large trees.

Kaarina: It's soothing, it's lovely, it's like a green corridor. There are owls and bats there at night which is great in a built-up area.

Melia: Birds sing there more than almost any other street that I have heard. There's that absolutely fantastic huge sycamore in the middle of the avenue. An amazing tree!

Kaarina: Which even the ITP [Independent Tree Panel] said 'exceptional measures' should be taken over.

Calvin: That's one of the handful of trees that had a special mention by the ITP.

Kaarina: That particular tree needs a bit of a publicity boost actually, because it is gorgeous. But it does cause pavement disruption, which is acknowledged by us and the ITP, so it does need some support to point out that it is exceptional and worthy of special work to keep it. Basically we've had grief with Dronfield Tree Services, because they're the ones who tried so many times and finally got that tree, which was a horrible day for me because I was in a real rush, I had my daughter with me and she wasn't very well at the time. She's on the autism spectrum, she's highly anxious, and I was just going down to quickly check the tree. I had a message, or just noticed something was happening, I can't remember. When I got there, they were finally felling the tree. I couldn't hang around because my daughter was so stressed and I didn't have my phone, so I couldn't contact anybody. Some residents came out and just showed up at my shoulder. We couldn't do anything at that point and I remember the arb, Adam from Dronfield. I just wish I'd had my phone on me. He sat on top of the stump and crowed, literally crowed, and put his arms in the air and cheered, telling his colleagues to take a picture of him. It was just horrible. I couldn't stay because we were going to the shops, we'd just detoured. It was horrible to see a tree go that you'd defended so many times, to see it treated that way.

Melia: Wanker.

Kaarina: He's just an absolute wanker. So that tree went. And subsequently a lorry crashed into the wall behind it, which never would have happened if the tree was still there. They knocked it down, rather than have to rebuild the wall. Someone backed into it.

Melia: They took a patch of the cobbles away too! That was me who put it on Facebook, I was going for a run and I saw them working away and I thought 'I'll just run home with the dog'. I said I'll come back and ask them what they're going to do with them. When it went on Facebook that 'the cobbles are being taken' there was a lot of interest. Somebody put that a local councillor has been to talk to the workmen, it has attracted attention.

Calvin: So they came out to respond to the concerns about the cobbles.

Kaarina: We've got no tree there, but we've got the cobbles. That was a really emotional tree for me. It was so much defended. Since then there have been loads of effective actions on Spring Hill Road, beautifully effective actions.

Calvin: It's a bit different to the other streets that saw action in that it has mostly students living there. People who haven't been there very long.

Kaarina: We lost one tree with one supportive student living there. This was on the corner of School Road and he really got a lot of stick from his housemates. He was on his own in the house when the crews returned, and said he let the arbs sway him that the tree was going to damage the house.

Calvin: It's always tricky with any household with several people in it. I mean, we've had couples who have different views on felling. It's not always straightforward whether a house is onside or not.

***

Kaarina: Other streets in our area stay in my memory too. Brighton

Terrace Road was another 'Dronfield Tree Services' road. The first time I encountered them was at the bottom tree on the road where they actually blocked me from entering a garden I had permission for. Adam actually just stood in my way until I sort of managed to step in around him over a wall. And then it became a vendetta with Dronfield Tree Services, they would go back and forth between Mona Road and Brighton Terrace Road trying to get a tree whilst we had frantic messaging, 'They've gone! They're coming back to you!'. There were several occasions where they just went back and forth and one of those times they just left Mona Road without success and the next thing we know someone is saying, 'They are running down the road with a chainsaw running towards the tree'. We had a supporter there who was out of the house. Her daughter had rung her and told her what was happening. They got the tree that day.

Melia: I'd just gone and checked elsewhere and I was just coming round the corner. I could just hear people yelling and Greg [Walker] was just coming up, he got out of his car a few feet in front of me, and you could just hear people yelling, probably ten or fifteen of the neighbours out there verbally laying into the arbs. I just kept walking past, obviously I was late for the party on that one. There weren't any protesters that I recognised, they were just residents that were out there.

Kaarina: The lady who lived there had a daughter who suffers from anxiety as well. It's not the sort of person you want to be doing this sort of thing to. She looked out of the window and saw it happening.

Melia: They were just right in front of her with a running chainsaw. The tree wasn't even barriered off or anything.

Kaarina: No barriers because it was one of those 'hit and runs' and they were just so desperate. It became a game for Dronfield Tree Services, a dangerous one. I think they'd already got some branches in a previous attempt. I don't even think they got the whole tree down that day.

Melia: I think they had to stop, there was that much emotion and upset.

Kaarina: They got a couple of main branches, but it really did appear that he was running down the road with the chainsaw powered up and whether that's true or not I don't know, I didn't see it closely enough to be sure. However, even running down the road with a chainsaw out and no work zone set up...

Calvin: With Dronfield Tree Services we've seen an entire tree down, a full canopy, that landed almost on the other side of the road. I'd believe most things about them.

Kaarina: Dronfield Tree Services were mad in my opinion, crazy.

Calvin: They got thrown off the contract, and that took some doing given what was happening at the time across the city.

Kaarina: Sackville Road, too, had really mature beautiful trees. And one of them is gone now. Some dramatic things happened on the way though, that was when I was first pushed by an SIA goon. It was the first day that the bouncers appeared on Lismore Road and across the city, the same day. I just couldn't believe that he actually put his hand on me and pulled me back.

Melia: There were four of them that day. I went in the afternoon and the same SIA bouncer had his face covered.

Kaarina: This is the first time I'd seen the goons [SIAs] and I hadn't heard of them, I certainly wasn't expecting them.

Melia: They had no ID, nothing on them to say who they were.

Kaarina: I asked them who employed them and they wouldn't tell me.

Elizabeth: That was the day I found myself under a tree with my baby. Kaarina was walking swiftly up the road to get there because somebody had alerted us, and I was a few metres behind her. I turned my back and then Kaarina shouted a warning to me.

Kaarina: I was trying to gecko but he was pulling me away from the wall. I called Elizabeth's name out like, 'Help! Help!'. I thought, 'Oh my god, I can't believe that'. I was frightened because I hadn't expected to be manhandled. That hadn't happened before.

Elizabeth: I was a bit confused. Because it was when they were putting the small four barriered work zones around trees, so I was outside the barriers and Kaarina had been geckoing, they were putting another barrier outside this barrier and I just didn't move. While stuck there geckoing I started breastfeeding. I had my dog with me and I had my baby with me. It was all very nice, quiet and calm.

Calvin: They've done this thing where they create a maze of the barriers, so you don't know if you're in or out. They think you're in when they say so.

Elizabeth: The short security guy came and I was filming and wouldn't stop filming. He said to me at one point, 'You're at risk of harm', and I kind of repeated what he said, whilst still filming, because I didn't know whether he was threatening me, but he said it was unsafe because the arb was already up the tree. He had just climbed up the tree, he wasn't doing any work on it, he was just fastening a second rope or something.

Calvin: I wish I'd filmed more events during the campaign. It turned out to be so important. The evidence we have and the evidence we don't have.

Elizabeth: He was looking at me when he said it. The baby was in the buggy with the rain cover on, so if there were like bits of lichen falling off the tree it wasn't landing on him. I started filming again when he said that and he came and he asked, 'Are you going to leave?', and I just ignored him; I chose not to engage, shall I say. The arb came down. We moved out of the way so he could come down and not be in danger of the ropes swinging around. He was still hanging around, thinking as to what to do, and then baby wanted feeding so I just sat on the kerb under the tree and fed him. I think at that point they really wanted us to move. Somebody had got me one of the neck scarves, and I had toothache really bad, so I covered my face.

Kaarina: You looked like a real desperado!

Elizabeth: He said, 'There's no point covering your face, we know who you are!', because Kaarina had shouted my name out earlier! I didn't say anything to him, but the reason why my face was covered was because it was quite a cold day and I had toothache. I remember laughing when he came up to me and said, 'I have to be willing to physically remove you'. He didn't say, 'I will physically remove you' because he kind of stumbled over his words like he'd only just been instructed what to say. Next to the top tree, Susan [Ashworth] was there, we'd been at Spring Hill Road and Susan was geckoing. When I got up there, Chris [Rust] was there too, and some others. The felling crews were cutting and there was sawdust all over, and people were filming. There were loads of them, but nobody was stopping them. I left the pram, went through the right of way, from the next road and entered a garden close to the work zone.

Kaarina: They got grumpy because Elizabeth was there with the baby, but there were people who were even closer who they were chopping branches over the heads of; the hypocrisy of it.

Elizabeth: It's really me isn't it, targeting me. The arb who came down that tree, he threw that branch down. He was furious with me. I was on private property, I wasn't even on the public highway, and when he came down off that tree, I think three or four security guards had to hold him back. I was quite calm. He was almost flying at me. He was restrained for coming after me, when I went and sat on the wall next to Susan, Chris came and told me that the security and arbs, because I had to move quickly up onto the wall rather than slowly, thought I was going to dive over the barriers with my child, so they shouted when they saw me going to sit next to Susan.

Kaarina: Once again, the fact that Elizabeth had a baby with her was an issue for them. If they thought that was so dangerous, what about Susan? An adult woman they were felling branches over, they were happy to do that, but as soon as a baby appeared they were saying, 'I can't believe you'd endanger a baby', as if she would.

Calvin: There's a running theme in the campaign. Everyone that's brought their kids to a protest has had this sort of targeting, with threats of social services.

***

Kaarina: I have a story about Spring Hill Road that's not so dramatic in some ways; the whole Barrow Road corner tree is dramatic and horrible. I think it's more of a story of the beauty of the road itself. The fact that it's a tree-lined oasis in an area which doesn't have that many trees, and it has got a lot of locals, like Richard and Judith, he came out one day to help Greg and he just saw what was happening. He'd been informed, he's the old chap, and he just came out and Greg said he was amazing, he just went under the tree and stood his ground.

Melia: There's only one resident on that street who was openly against tree preservation.

Kaarina: We had some very supportive student households. And the story was about cohesion, and the value of such a road in an area that doesn't have that many street trees. It's really unusual to take eleven out of seventeen on this road. I've heard students saying, 'You know, the reason I picked a house around the corner is because I get to walk down here every day'.

Elizabeth Mountain: That day turned out interesting for me! What happened was that the barriers were up, and I must have argued about fifteen minutes for permission to get into a property where we had a garden permission. When they did eventually move the barrier, I put my baby at the end of those steps, we were just in front of the property, where the gate is. Well I just stood here and did my 'geckoing' thing because there were already two people under the tree in the garden. But by then baby had got a little bit cranky so I picked him up, and I've got dodgy joints and I had to straighten my arms.

Calvin: So you're being pushed by the security guys whilst looking after an unhappy child?

Elizabeth: The security guy just a few feet away from me was pushing the barrier up against me here and the baby, he was throwing himself back, right down so he was in front of the barrier; not one of the security arbs did anything about the fact that he was in front of it. And then I mentioned that if I'm holding my baby over the barrier shouldn't you be stopping work? That was the bit that was filmed. I was asking them the question.

Simon Crump: So you messed up because you had to dangle the baby in order to straighten your arms!

Elizabeth: I'm not Michael Jackson! Baby was throwing himself back and being a bit cranky. And my arm was locked in and was getting stiff so I straightened it out. The barrier was in front of me so he was technically over the barrier.

Kaarina: Never mind that they were bashing barriers against you.

Elizabeth: I've been threatened several times with social services. The SIA bloke, 'Uppercut Guy' – he's threatened to report me for having my baby with me on protests. I did say to them they don't frighten me or intimidate me when they threaten to call social services. I said, 'Oh you can have my social worker's number. She's called Linda.' Don't get me wrong, it's not child welfare services, it's because I'm autistic and I had agency support at the time. I have a letter from Linda that says she encourages me to take positive risks in life!

2017-2018

'Then on that day in December 2017 the vehicles came down the road in near silence, engines switched off. It was the flashing orange lights that woke us up.'

Sue, Resident

Vainor Road is a residential street in the Wadsley area of the city.

Sue: The first I knew of the policy, and how it affected us on Vainor Road, was when I saw Christine [King] and Rebecca [Hammond] putting up notices on the road. Then we received the ITP Survey in the junk mail envelope. Soon after that vitriolic newcomers to the road were loudly supporting the planned fellings, and criticised the rest of residents for wanting to keep our trees. I was elated by the survey result, which resulted in the majority of households who responded voting to reject the Council's plans, until it became clear that the Council were going to ignore the Independent Tree Panel findings that most of the trees should be retained.

Calvin Payne: In December 2016 I was at the full Council meeting. It was a couple of weeks after Rustlings Road and a month after our arrests on

Marden Road. Amongst the petitions was one calling for the removal of trees on Vainor Road with twenty-two signatures, but without anyone there to present it. It was all very strange, the road had voted to go to the Independent Tree Panel (ITP), and the majority who replied to the survey were against the Council plans. It was suspected that local Labour councillors were behind the petition, and officially the reply to the petition from Bryan Lodge was that it 'showed that there were different opinions with regard to highway trees'. Something which none of us doubted or denied.

Sue: Nobody asked us, or anyone else that opposed their policy, whether we were in favour of saving the trees. The first we heard that there was a petition about our road was when it was presented. We knew we had won the survey vote, and up till then we'd had no response from local councillors at all, despite many emails and phone calls. They just weren't interested.

Christine King: Our group spent a lot of time in Wadsley, around Vainor Road and surrounding streets. That area was hit hard before Rivelin Valley Road was.

Sue: Bob Johnson [local ward Councillor who became Leader of the Council in January 2021] was very active on the STAG Facebook page at the time. He accused me of chopping down mature trees in my own garden! Rebecca told him that he was basically stalking me now. I advised him that his informants were misleading him and that I've been planting trees in my garden, not felling them. The only trees we had felled were conifers over thirty years ago which were encroaching on a neighbour's garden! I thought that as the Council was a public sector organisation, that when the road went to the ITP that decisions would be made properly – open and above board. But when you look back at all that was said and done by people like Bryan Lodge, Paul Billington and Jack Scott [former SCC Cabinet member], they were very deceptive and completely untrue weren't they? In my job governance and openness is my responsibility, and I'm really clear to people about their public duty, adhering to the Nolan Principles for example, so when your local authority acts in this way it's just incredible and quite unbelievable.

Siobhan O'Malley: The first time I seriously tried to stop a felling was before the injunction on Vainor Road. I didn't have a fucking clue what I was doing! I'd just read a few things on the STAG Facebook page. I'd argued with felling crews before in the area for chopping down healthy trees, so by the time they got to Vainor Road I was like, 'Fuck this!'. It was a massive tree and there was no-one there. I thought, 'This is ridiculous', and I just stepped into the work zone. The arbs were really annoyed! It was lawful and peaceful and I just did it.

Sue: The first time my husband and I were there on our own when the felling crews turned up I didn't really know what I was doing, I just desperately wanted to save the tree. My husband was home and he called me at work. I managed to take the afternoon off and came home. I pulled up and ran down the road asking for our neighbours to come and support us. Some came out, but others were nervous about getting into the barriers or getting tangled up with the abusive neighbours. People weren't really sure what they could do. When we demanded that the crews check for bats one antagonistic resident shouted out, 'There are no bats, just that batty woman!', pointing at me. I went to Milden Road just round the corner because they were setting up there ready to take the tree at the end of the road. Rebecca was at the other end doing the same. The arbs were really friendly and just said that they would go. They were fine with us then. There were evidence gatherers there too, taking photographs. I'm sure we are there somewhere as part of the 'wanted squad'! As the arbs left, the 'handyman' from the house at the end of Milden Road started watering the garden, but was intent on hosing us standing in the barriers. One woman walked past and shouted [at us], 'Can't believe you're doing this, I thought you had more sense!'.

Geoff: There was a time when all the angry residents gathered together. They saw us and made a beeline for us. 'Gandalf' pacified them a bit, but there was a small, vocal, loud group of residents who were very unpleasant.

Siobhan: Peter [Giles, aka 'Gandalf'] got a lot of stick from one woman. He was by the tree next to her house and she came out and shouted that he 'looked like a paedophile', all sorts of nasty things.

Sue: As Peter remained under the tree outside the woman's house a builder started cutting paving slabs, making sure that the dust went all over Peter – he must have been breathing in the dust.

Christine: I think that was the same woman who said we looked like 'a sack of perverts'.

Sue: When the felling crews first came here one morning, one woman in particular was being abusive, she told Siobhan that she 'looked like a prostitute', and all sorts of other unpleasant things.

Christine: The crews started coming out earlier and earlier, until we were seeing them off at 6am. The next day they arrived ten minutes earlier and so did we! I slept in a supporter's spare room for a week on Milden Road to ensure I was there first.

Sue: I'd seen Christine and Co at six o'clock in the morning standing by trees, freezing cold, and they'd stay the whole day under there. They were amazing!

Rebecca Hammond: Later I was back on Milden Road outside the house of the friend of the unpleasant woman on Vainor Road. The thing was that I had just been sent the pre-injunction 'undertaking' by the council [just before the July 2017 injunction hearing] and I planned to sign and return it the next day. I thought, 'Well, this may be my last opportunity to do anything'. I went into the barriers, stopped the work, and she turned the hosepipe on me. The first time it was fairly brief, just a sprinkling really. I moved to the far end of the work zone, away from the arbs and away from the woman standing in her garden holding the hosepipe. She started arguing with a neighbour who was on our side. A Council evidence gatherer was there winding her up, and I pointed out to him that he was there to gather evidence, not to wind up residents. She threatened me again and I threw my handbag to another protester, which was just as well as the woman turned the hose pipe right on me, on full power. I was absolutely soaked. It was a warm day, but I was soon turning cold. I considered it an assault and called the police. She told them that she was watering her garden and it was an accident, but we had it on film, which showed a different story. A couple of weeks later I got a sort of apology from her in which she admitted that she had got me 'slightly wetter than planned'!

Sue: I recall one day after collecting our granddaughter from school during the summer when she quite innocently asked why people were standing around the trees. I remember her being truly upset at seeing a baby bird at the foot of a fallen tree which had obviously had a nest in it. She was heartbroken and we tried to explain what was happening, and she said, 'But what about all the other creatures, Grandma, where are they going to go?'. It was just so sad. Other local children had given trees names – 'Lulu' was a favourite. There were several road closures when nothing happened, but we had cars parked under trees, just in case. I often spoke to Siobhan, Christine, Geoff, Rebecca and Peter during those times and we seemed a bit of a lonesome posse sometimes, not like Nether Edge or other places. I think by then we thought they'd forgotten about us, that they'd go away and leave us alone, after all they had resurfaced the road and the footpath so there was no reason for them to come back and take our trees. They had taken trees on surrounding streets and it seemed had left us until last, but now I think we probably wound them up, and the more we defended a tree the more they were determined to take our trees – it was almost like a personal challenge for them and they were determined to win no matter what the consequences.

Geoff: Towards the end of 2017 they got the road closure again for Vainor Road. We were there early one morning with three trees left on the road. I saw the crew go past me and target another tree further along. Three of us were trying to protect three trees while also trying to protect each other.

Sue: Then on that day in December 2017 the vehicles came down the road in near silence, engines switched off. It was the flashing orange lights that woke us up. Then the SIAs arrived, people in hi-vis jackets – it seemed as if there were dozens and dozens of them, trucks full of barriers, cherry pickers. I looked out of the bedroom window and saw the doubled up Heras barriers outside to try and stop us from bounding over the fences to get to the trees – I just couldn't stay and watch. I asked my husband to drive us off for the day, I just couldn't bear to see the carnage, or to hear the sound of the chainsaws. Unfortunately whilst in the car we heard the dulcet tones of Julie Dore, Leader of the Council, on the radio defending having people sent to prison for breaching the injunction. At around 5pm I phoned a neighbour to ask, 'Have they done it?'.

Christine: It was like an industrial site, doubled up Heras barriers placed on top of each other to block off access. One of the trees had been singled out by the Independent Tree Panel as special, *'a prominent specimen worthy of exceptional measures'*, and despite that they put all that time and effort into felling a tree singled out for saving.

Sue: Not only did we lose the trees, we lost the community as well. People don't talk to each other anymore, there's definitely an atmosphere on Vainor Road now. I wonder if the councillors had thought about that. What happened has marred what was a lovely road, with people not talking, relationships ruined. Whatever party people represent they have a duty to be honest, open, transparent and fair to the public, but Bob Johnson just set people against each other, that was clear from the interaction on Facebook a long time ago. I have mixed emotions now. I'm proud of playing a part in what the campaign did. I'm genuinely sorry for the young people who won't know what they've lost. A lot of the bad feeling still remains. We still live with the same antagonistic neighbours, people who believe that cracks in their house were caused by the trees, or moaning about having to clear up bloody leaves. None of this should have happened and if the Council had done their job properly – been open, transparent and above board – there would have been no need for any of this.

2017-2018

'It's not too late to change your job. You're a young man. Why are you in tree surgery if you don't like trees?'

'Gandalf' (Peter Giles), Protester

Rivelin Valley Road is a main arterial road into Sheffield. It was built in the early 20th century and lined with lime trees.

Rebecca Hammond: I remember the events on Bannerdale Road in June 2016 and the online discussions about the implications of direct action and the police attending. That was the first piece of concerted non-violent direct action in the campaign that I had seen or knew about. It was a bit of a turning point. I think it was when people realised that 'this is how we can save trees'. The Rivelin Valley group held a NVDA training event, organised by Fionn [Stevenson]. I'd been to one that she had run elsewhere, and then we organised one ourselves. I think people were still quite scared but there was an awareness of the need for it. After that, we organised a call-out phone network.

Calvin Payne: The Rivelin Valley group was similar to the one in Nether

Edge in that it was formed in 2015 when the attempts to fell trees in the area were still some time in the future. Other parts of the city didn't have the same luxury of time to prepare themselves of course.

Rebecca: I woke up to the Rustlings Road events and was absolutely incensed by that. I phoned Radio Sheffield later that morning, crouched behind the bins at work! It was the first time I had been on the radio. I just felt I had to hold myself together and get the message across that day. Not long after that, I presented the Rivelin Valley petition at the monthly full Council meeting. I knew how crap it was at the Council meetings, but to be personally on the receiving end of it...well. I remember writing about it on Facebook that night. It was shocking. We tried the 'right' ways of making our views known, and in the end you just say, 'We'll just go and stand under a tree, then'. If there was an easier way, we'd have done it. That winter we were expecting attempts on Rivelin Valley Road but it didn't happen. I went to Chippinghouse Road in February 2017 to give my support, to show my face. It was awful, freezing and depressing. I don't know how people stuck it out all week.

Calvin: I preferred absolutely anything to attending Council meetings. Asking a question or presenting a petition and getting a reaction every time which ranged from disinterest to abuse. Horrible, shameful, pathetic. In public as well, in front of the media.

Rebecca: I can't remember my first direct action. It would have been at a time between the Trade Union law and the injunction. I used to do a quick patrol or drive round a few streets on the way to work. I had got complacent as I had never seen anyone, then one morning there was a chipper and a felling crew! I parked up and sat there. I didn't engage with them, and they came and asked what I was doing, so I said, 'I'm waiting for you to leave'. I posted on Facebook and thought to myself, 'I'll have to be late for work then!'. It was a bit surreal, they were watching me watching them. I didn't think anyone on the other side knew me at that time, but later the incident was in the pre-injunction papers with me correctly identified.

Geoff: The first time I turned out I saw Christine down there, gathering the forces. She seemed to be co-ordinating things.

Christine King: I think I had pink hair! What made anyone think, 'I've never been at a protest. I've just arrived. I need to find a responsible adult. Yeah, the one in the pink wig!'. I wasn't expecting that, I don't think. To be fair, that was why I was wearing it, not just as a disguise, but to bring people to me so I could explain the injunction to them. I thought, 'Break it or don't break it, but know which you're doing'. I couldn't tell you how many times I explained what 'Persons Unknown' meant.

Rebecca: That summer, before the injunction, we had evening patrols when there were night-time road closures for resurfacing. One evening, I saw an Acorn truck on my way there. I quickly drove to Rivelin Valley and, oh shit! – they'd coned off one of the trees. I had that shaky moment when I needed to alert the call-out group. I was trying to decide which would be easiest – text or call? I stepped inside the barriers. I was probably only on my own for ten minutes, but it was the longest ten minutes of my life. Very quickly, people started to arrive, one, two, then lots. Although plenty of people were there in support, I was the only one inside the work zone. I was just standing around and getting a bit cold and the message went out on the call-out phone, 'Can someone bring Rebecca a jumper please?'. It was such a relief when people started to arrive. Having support outside the barriers was so important.

Christine: One of the first trees on the road they were wanting to cut down was one where a resident had been told the tree was dangerous, and so she was quite aggressive about it coming down. She'd been abusive to us for probably two years. Every time anybody went out she was saying things like, 'You can't stand on the pavement. You can't stand here! I'm phoning the police!'. She'd been completely abusive to us for years, but the tree did have a big lean on it, and a local tree campaigner had said to me, 'To be honest, if that tree was outside my house I wouldn't be very happy'. The group decided they weren't going to contest it, even though she'd been spectacularly nasty. If you went to bed every night afraid that a tree was going to fall on your house, well, that's not great, so I decided I was going to be the better person and not protest it. And at the end of it, when they'd cut it to a trunk, the arb was being showy but he was really wobbling it about, and it was really wobbling. I think the road crew had damaged so many roots that it was unstable. I wondered if it had been like that before the road was done, because we've had that in other areas where residents had said the tree was alright, but now after the road was done the tree wobbled like anything, so actually it may have needed to come out. I remember Russell [Johnson] coming up and asking, 'Are we contesting this tree?'. And I said, 'Well, no we're not'. The next day, the same crew was at another tree, the barriers were up, and I was thinking, 'Oh shit!', and then this masked person just appeared and went, 'Boing!', up and over. I just looked and thought 'Awesome!'. It was my first experience of 'Running Man'. He warned me that people were being followed home. He sent me a message at the end of the day that said, 'Well, I was followed home'. He told me they couldn't keep up with him, and I called him 'Running Man' because he legged it through the woods to get away. He had this complicated thing where he did a lot of shopping in charity shops and then he had the arriving disguise, the bunnying disguise and the going-away disguise, so he had carrier bags strewn around the woods.

Siobhan O'Malley: Wow, Running Man did some serious prep! I had seen it come up on the WhatsApp group that people were being followed home by the evidence gatherers. I looked around and there was someone in hi-vis somewhere on the way up the hill overlooking the road.

Calvin: I think I drove down there with Alan [Simpson] and they were filming us from the off, as we parked about a hundred yards further down Rivelin Valley Road. Alan was dressed as an evidence gatherer: hi-vis, cardboard camera, the lot!

Siobhan: That was genius, that. Very funny!

Calvin: He was nicked after about two minutes. It was a great idea, but he spent the day in the police station. I don't know if impersonating an evidence gatherer is an offence! The annoying little squit, the little dark-haired evidence gatherer, accused Alan of assaulting him by putting a cardboard camera in his face.

Christine: It was Wednesday and Geoff sent me a message saying the felling crews were there. I remember driving down the road, I drove past them and I thought, 'Oh shit!'. I parked in the fishing-lake car park, threw a disguise on, and ran back down the road. I waited until I could cross the road without getting run over, and climbed over the barriers.

Geoff: More than once I was struck by Christine's bravery, my first experience pre-injunction of seeing someone go inside a work zone. She'd march over, go straight over the barriers, and that was that – job done! That was remarkable to me. It was because of those actions that I got a bit more involved in the campaign. Seeing people who were willing to go out on a limb and potentially put their freedom at stake.

Calvin: People knew the consequences only too well. They were prepared to take the risk for something that mattered.

Siobhan: I turned up once, fully expecting to go into the barriers on my own, and then there was this chirpy jolly guy with a bandana covering his face saying, 'Are we getting in, then?'. I was like, 'Yeah! Let's do it!'. And over the barriers I went. I didn't think they'd be that good at recognising people. I had most of my face covered but not my eyes. You can sense danger. That's the thing isn't it? All of us can sense danger. You have to go against every instinct when you know there is this aggressive attitude towards you, and it does make it easier if you're incognito, doesn't it? Just a bit.

Christine: I sort of psyched myself up to do it and hide my identity. It's why I needed to be there first. Because I can't get past them. It's too frightening. If I was braver, I'd have done it more.

Calvin: Plenty of people around that time were saying that they were prepared to break the injunction if they were in disguise, for various reasons. There were some with totally different personas, quiet outside the barriers, but just extraordinary with their masks on, totally unrecognisable.

Simon Crump: There's probably another side to it for the security guys in that it's probably a lot easier to beat somebody up if you can't see their face.

Geoff: They also know that there is going to be very little legal comeback, because people can't expose their own identity in order to report someone.

Calvin: People were being rewarded by Amey and the Council for doing that. When we were in court, the 'Uppercut' SIA man was being paraded by Paul Billington, smirking at us sitting in the gallery. He's well regarded by the Council. He did his job for them.

Siobhan: It's so sick! How are we supposed to live in this city when that is the overriding attitude of the people in charge?

Christine: They were noticeably more polite to me when I turned up one morning in Norton Lees and they didn't know what gender I was. They were discussing it. At one point they said, 'Is it Paul Brooke?'. Well, actually I think that I'm a bit taller! Then they just left me there for about seven hours, which was a bit tiring. I was asked to go to Thorpe House Rise and I thought, 'This is going to be a mistake', because the SIA let me out of the work zone, and then of course later, I had to get back in. I'd already worked out there was a way under the barriers, so I snuck in. I stood there for quite a while, and they didn't even realise I was there.

***

Geoff: There were other things that happened on Rivelin Valley Road; there was one day in particular when there were loads of extra security, observation guys, and evidence gatherers. I was going out about 8 o'clock in the morning, just to see if they were turning up. I was hanging around Mousehole Bridge and there was a police officer on the other side, with two other people starting to walk away, and because I'd only just come on board

I wasn't sure whether they were with me or against me. I just kind of eyed them up. And it transpired later that they were also evidence gatherers. It was a bit like a spy film.

Siobhan: I filmed them filming us on that day. That was the day I dropped by briefly but I couldn't stay for the whole day. So I came to film a bit and just add to the crowds. I had to turn away because they started shouting my name at me.

Geoff: Then the felling crew turned up and I texted Christine, and they were ready to go, but they didn't start cutting, and one of the evidence gatherers went across and had a word with the arb, Jason Wignell, and he just sort of stood down for a while and I really freaked out. I didn't know what was going on and I ended up going away because I was afraid that I was being set up. It was obvious they were setting up for something. It transpired later from the Freedom of Information request, that Rebecca made, that it was the day they were trying to get photographs for the *Star* [local newspaper]. It was all a set-up.

Christine: I had a doctor's appointment that day. Every other day I just walked up to the cafe and thought, 'If anybody follows me they're going to have a lot of staying power, because I've just basically come for a walk up the valley'. But that day I couldn't afford to spend an hour shaking somebody off.

Calvin: That's the thing about direct action: you've got to be free for the day because you don't know what you're getting into. You don't know how long you will be needed or stuck somewhere.

Siobhan: I was taken to court about one of those days on Rivelin Valley Road. There were already a few people there, but nothing had been properly set up yet. There was no fencing up. It was just starting to be unloaded, and yet there were a number of people there, and then a bunny I've never met before marched up. I had been wanting to intervene because the Rivelin Valley is a historic and beautiful place. So many memories for me and my son. So when I saw what they were doing, I was just preparing myself. I'd gone down there with a scarf on my face. I knew that they were watching us and following people, that sort of thing. I put my scarf over my face, put my hood up and got down there. We were full of enthusiasm and really excited to just sort of say 'No!' to the felling. The bunny chap, who I'd never met before, was really bouncy and I was willing to wait and stand back and see what level I was going to let them get to, a sort of level where they were clearly doing something bad and then jump in. But he was saying, 'I'm going in', and I'm thinking, 'OK, well I'm going in too!'. You've got to do things with 'safety in numbers' in mind, that sort of thing. We just hopped in and proceeded to have a nice little chat together about nature, all kinds of lovely stuff. I learned so much stuff talking to that guy.

Calvin: It is certainly easier stuck in that situation when you can have a chat about something else to pass the time!

Siobhan: At the time I was just thinking he was this lovely jolly chap who obviously knows a bit more about this sort of thing than most. We sang a verse or two of that Everly Pregnant Brothers song, 'Don't Cut Me Down'. It seemed quite amicable at first, but then the evidence gatherer guy came over, the short one with the dark hair, and he started threatening us with arrest and all sorts of other things while filming us. And then it started to get a bit funny after that because, well, we'd been trying to talk to the arbs a couple of times that week, and one of them had been incredibly abusive. The other protesters would be talking to them and I'd try to pitch in something equally valid. I'd just get him picking on me going, 'You're the kind of protester I hate actually. It's pathetic'. I was saying, 'I'm not even an activist. I don't know who you think I am, or what you think I am. You've obviously got me tagged as some sort of mental hippy chick or something'. I remember what I said that triggered him, and I still remember it vividly because I think it was a salient point. He was saying, 'Well, our company has done their expert checks on these trees and said they need to come down'. I told him, 'Your company experts are biased, their views are going to be filtered by the priorities of your company'. That was what made him start having a go at me. Pointing out what I thought was obvious about

Amey's interests shouldn't have got me a mouthful of abuse. They did eventually pack up. They hadn't fully set up the barriers anyway: it's well known now, they set it up in a sort of 'C' shape, using the wall basically as part of the barrier. Even that wasn't complete because it was going straight into somebody's drive. They were just walking in and out of the house because you know the residents were pro-felling, tree haters or whatever. Serial sweepers I call them because they're obsessed with leaves. I didn't realise that anything untoward could happen from here because it had been such a slipshod attempt, I couldn't imagine that they could get me for it. At the time I was sort of, 'Yeah whatever, I'll take it, try and take me to court', but I obviously did not think that it would be physically possible for them to think anything other than, 'OK, fair dos, you've intervened, it's not a great time for us to be doing it'. I've got footage of them dropping sections of a trunk that must have weighed half a ton whilst cars were passing by. I thought it would be madness for them to suggest that I was the one doing something, anything, wrong.

***

Andrea Stone: One morning I saw the call-out message about Rivelin Valley Road. My childhood was spent in the area and it was important to me. My previous protests had been around Nether Edge, and this was my first time going further afield. I had been arrested in the past and used to go out prepared. I knew that if the tree-felling went on I would get arrested, and I was prepared for it that day.

Siobhan: The next day they were back and so were we. We all hung around, they'd mostly packed up, and then I set off home because my son had turned up at this point, I think. We went up the hill, it was only then that I saw there were some hi-vis lurking around Malin Bridge and Holme Lane. I didn't see any cameras but they must have been taking pictures of me, because they had a photograph in the court evidence with my scarf having been pulled down a bit. I don't know if they followed me all the way up the hill, but I think that would have been very brave of them because it's fucking steep. They're not paid that well. They traced me on Instagram and

Facebook. They found out where I live because it's a Council flat. As soon as they got my name they could just put me into their file and see where I live. That's what I got for being a Sheffield Council tenant.

Andrea: I'd been on Kenwood Road and Thornsett Road before, stood out in the snow. That is in my local area and I had taken to setting off my personal attack alarm at tree protests as a disruptive nuisance, but the batteries would run down too quickly and I thought I needed something else. I'd taken a plastic trumpet out a couple of times, but it had stayed in my bag. I had decided that if I couldn't stop the felling I was going to make as much fuss as possible as well as being as disruptive as possible to their operation! On other demonstrations friends had moaned about it – so I know how annoying it is. It's a GMB Union horn. It is for making noise on demonstrations and protests. It's years old now, and whenever I've been on protests it's been with me.

Geoff: Rivelin Valley is sort of my patch. In the New Year, I thought they'd come at night as they had started doing in other parts of Sheffield. I started camping out really early, two o'clock in the morning, in my car. I didn't really feel comfortable hanging around the houses, so I parked near the café. Also, I thought it was more likely they'd come to the trees near there. Even though officially they weren't going to start until 8th January, I was there waiting during the week before. I was also there on the 8th, and that's when it started going strange. On the morning of the 9th, that was a Tuesday, I planned to pack up around 5am but I decided to hang around the café a bit longer. Then I set off for home at about 5:20am. When I got to the fire station end, it was 'Oh dear, no!'. The crews had arrived *en masse*. Basically there was a hundred yards' stretch of road that was roped off with red and white ribbons and cones, and there must have been at least ten workers, including four arbs. Other campaigners had been watching this end of the road during the night too. I parked up and just filmed everything and I got half an hour of footage. Two of the arbs were still up in the trees, and I was standing underneath the trees, encouraging them to come down. There was an air of guilt about them because they hardly spoke. And once I arrived, it was obvious they were going to have to tidy up. So they spent the next twenty minutes tidying up. It was really dangerous as well, what they'd done. There were no road signs. Rivelin Valley Road is a really busy road. The cones were right out, almost in the middle of the road.

Gandalf: We had months of sleep-deprivation, blizzards, rain, frosty mornings, anger and frustration. I've experienced violence, unnecessary arrest and being reported for summons, and all because I am seen as a 'tree-hugger'. I've stood defiantly for hours through many seasons against increasing numbers of workers in hi-vis jackets who were hell-bent on needless felling.

Geoff: Two mornings later, I was out early again, expecting them to show up. This time it was Jason Wignell and two others. They were up one tree when I arrived. As I approached they said, 'No, no, stay there!', and it was because there was a branch just about to come down. The arb said he'd come down once the branch is down. He pushed the branch and it fell way outside the work zone, the so-called safety area, right across the pavement. They packed up quickly and drove off after that and I thought I'd do my James Bond stuff and I followed them up Rivelin Valley Road. I just wanted to see where they were going to go next. There's a little road just before it joins Manchester Road, which I turned onto, and as I did so there was no sign of them. I knew instantly what they'd done – they'd pulled into the car park there and I thought, 'They've pulled into there, trying to hide, they've got their lights off'. I thought, 'Oh well, it's a losing game here', so I parked opposite. I decided to sit here for a while until they came out. And so I sat there for ten minutes. Nothing was happening and I thought, 'Well I'm getting a bit bored now, I'm going to provoke them into something'. I drove back towards the road and shone the light briefly in their direction. They emerged like bats out of hell, they came charging out. It was Jason Wignell driving. He deliberately swerved towards my car as if he was going to crash into it. I turned right onto Manchester Road, he came around the other way, again we sort of crossed and again he swerved toward me deliberately, and I just waved them goodbye.

Siobhan: They did that to me and my ex when we were looking out for them when they were doing the night road closures. We were saying we'd all be dead by now if we lived in a country with more fascist laws.

Geoff: I drove back towards where I'd camped outside the café as I knew someone else would be turning up. When they arrived I suggested they go down to the fire station and let me know what was happening. As soon as they got to the fire station the Amey crews arrived again. They'd gone back the other way. Just past the pedestrian crossing Jason Wignell stepped out in front of my car, which made me stop abruptly. I tried to reverse so I could get past him and he just followed me and stood in front of the car again. I couldn't really reverse because it was a very dark corner and I kind of got myself into an awkward position against their vehicle, which was in the zigzag zones of the crossing. He had his young assistant with him and the two of them did a slow march in front of my car, copying what we enjoyed doing to them. Another of our protectors was parked up, and I managed to message her and she came along and filmed what was happening. By now, one of the arbs was up the tree, trying to attack it, and one way or another I was able to gecko and get them to stop. Then they packed up and went away, but then they came back again a bit later, just in case I'd cleared off. I was a bit shaken at the end of all of this. It had all been on my own, or with just one other person.

Christine: Two other protesters messaged to say they were going to come over, they were going to come again because the arbs had been horrible, so they might think they've seen us off. They were going to come over for half past four and I asked my companion, 'When you come round what do you want to do?' Because we knew they were going to be up the tree as early as they could. So we agreed to be out at half past three. I picked them up because they don't live far away from me. We were parked up on the right side of the road; I was pointing upstream when this blue van went past. 'That's our cue', I said, so we leapt in the car, followed, and it was Dave Smith [arb superviser]. He parked up at the café so I parked behind him.

Calvin: They weren't doing the night raids for long, were they? Although it seemed like it. They were definitely losing their rag. They were thinking, 'If these people can stop us at four in the morning, how can we beat them?'. I'm pretty sure they got to a point where they are thinking, 'We can't even go out at this time without them stopping us'. And the safety of working at that time, they can't possibly think that's OK. They knew it was unsafe and crazy.

Christine: Gandalf was there as well; he was talking to an arb, Dave Smith, through the truck window. 'It's not too late to change your job! You're a young man! Why are you in tree surgery if you don't like trees?', he said. They did try to get up a tree at the café and it all got into a bit of a kerfuffle, the police were called and it all got a bit difficult as we were all spread out along the road defending two groups of trees almost a mile apart. They'd just been driving up and down Rivelin Valley Road and not getting anywhere. There was a point when I walked down the road, I was by the three most pruned trees at the fire station, and Jason Wignell was doing his loop. He turned up and he was on the other side of the road. He saw me, so he came across to my side of the road so I stepped out between the kerb and the carriageway so I was on the road and off the road at the same time. He screeched across, stopped in front of me, and then he was doing this thing in his big vehicle where he just kept coming towards me and stopping, coming and stopping. Coming and stopping like he's saying, 'I'm going to run you over with my big penis extender', going to run into me. I'd been up for around four hours by then, cold and tired. I'd had enough. It annoyed me, so I just turned my back on him; he swerved around me and drove off. The thing that annoyed me most was that my friend got run over a few months before and she still has a huge metal frame on her leg; she's going to have it on for months yet. I was thinking, 'If this dick runs me over, am I going to have the same broken leg. Will we make a decent pair of bookends?'.

Siobhan: I think their problem was they had so much contempt for us members of the public, they just thought we were a load of idiots who wouldn't notice what was happening under our own noses. To actually see us all showing solidarity and being able to go out, express our opinions and stop them from doing what was wrong, they didn't like it.

Calvin: The incident on Rivelin Valley Road that got the most attention towards the end was 'Tootgate'. Throughout the campaign our actions were often a mix of stunts and serious hard work, but all contributed to delaying the tree-felling, which was sometimes all we could do, and to showing the other side that they didn't have it all their own way. Sometimes we just had to be annoying!

Andrea: When I arrived I could see that I was too late. They'd already got the barriers up. I walked up to the fence, got the 'tooter' out straight away and went 'Waaaah' as loud as I could! There were other protesters who disagreed with me doing that, which I responded to in no uncertain terms: 'If those people are chopping down trees, or facilitating that, then they're going to put up with my noise!'. I blew it a couple of times, the arbs put in their ear plugs, and then I saw a commotion going on further up the road. Christine was on the floor, and there was a scuffle going on as I walked towards the area. A resident came over and said to me, 'Haven't you got anything better to do?'. I think it was his complaint that led to my arrest. I carried on tooting as I walked down the road. I could see that people were annoyed with me and I thought, 'It's working'. A police officer came up to me and said, 'There's been a complaint'. From being arrested years before for causing 'nuisance and distress' I knew the score. I knew I couldn't have caused that to an arb, so I guessed it was a resident who complained. I didn't think about it, I just tooted loudly. It was basically a 'fuck off'. When the officer said, 'Do that one more time and I'll arrest you', I knew that's what was going to happen. He turned and started walking away. I thought, 'Am I going to let him shut me up? Am I doing anything wrong?'. I decided that I wasn't on either count and I knew what to do. Come and get me! As I was being led away the guy who had complained was clapping, so I carried on tooting, at which the officer snatched it away, breaking it in two. So I set my attack alarm off! I shouted all the way down to the police van. I was definitely on a mission. I wasn't there to have fun or just make a noise. I knew what I was doing all the way through.

Gandalf: I've met gentle folk who have become incensed with the Council's vandalism of beautiful, majestic, trees to the point that they have done things which the law construed as being illegal. People who stood proudly, sometimes alone, who gained collective strength from like-minded friends, who were prepared to stand up against what they believed was wrong. Although I shared the victories and sorrows of others, across Sheffield, my own troop of heroes was based on Rivelin Valley Road. Early battles were won and lost, and lessons were learned. The tide of hope sometimes began to ebb, but with every knockdown people got up and fought on.

2017 - 2018.

'A very fine specimen, in excellent condition, with a further 150 year life expectancy.'

Independent Tree Panel

Vernon Road is a residential street in the Dore area of the city. It has one street tree which is well known as the 'Vernon Oak', at least 150 years old, but maybe considerably older, and believed to have been a boundary marker before the road was laid out.

Margaret Peart: I remember that towards the end of 2016 I began seeing these horrid ribbons on the trees! I thought, 'Those look a mess'. I'd seen the ribbons on miserable-looking little trees in the area and thought, 'Fair enough, they can be replaced with better trees'. I never thought they would come for a tree like Vernon.

Ann Anderson: I'd become aware of the TPO [Tree Protection Order] applications being made by Chris [Rust] for Vernon and some other trees. I met a few people from the campaign, and before I knew much about it I'd been dragged into STAG! I thought the campaign was more developed than it was at the time, but everyone was finding their feet and in a similar position to me.

Sally Goldsmith: Early on, I remember getting a message from a friend saying, 'They are coming to fell trees on my road. Shall I park under one of them?' to which I replied, 'That sounds like a good idea!'. Not that I knew much about such tactics at the time.

Ann: On Totley Brook Road, when we thought the TPO decision was still outstanding, the felling crews arrived suddenly and trees were lost. I think now, 'What we would have done later on in the campaign...' – that was such an upsetting day.

Calvin Payne: Following the award of the High Court injunction in August 2017, Vernon Road was one of the roads that were scheduled for felling work. A road closure notice about the 'tree replacement' work was posted on the road and delivered to residents. There was quite a turnout from supporters and the media, with the BBC and ITV already present, early on Monday morning.

Sue Unwin: I was meditating under the tree about half past six in the morning when I was vaguely aware of a tall, slim, figure walking past. When I came out of my meditation I looked up and there was a man up in the tree!

Margaret: I was looking at my phone waiting for 7am, when the road closure started, and checking if there were any updates, when I looked up and there was a person up the tree. It was quite a surprise to us!

Calvin: With the media present, the unknown man up the tree caught the imagination. The strange thing was that the Council 'evidence gatherers' were also there filming and photographing the people that were just milling around waiting for something to happen. We had been used to their presence at felling sites, but had been told they were only interested in protesters who were potentially breaching the injunction. We knew that wasn't true, but it was more obvious than ever that day because there were no felling crews at the site.

Margaret: The reason given for felling Vernon kept changing – three times over the months. Trying to find out the reasons for decisions was very hard, impossible sometimes, which all added to the mistrust.

Tree Climber (reproduced courtesy of Sheffield Live TV): What am I doing up here? Well, this is a great opportunity to come and buzz up a tree! Secondly, I'm here and everyone's here because we think the Council could better spend their time, their efforts and their money on better things than trying to remove beautiful trees like this. They need to be better at listening to us, and this seems to be the only way we can communicate what we want. I'll be here for as long as I need to be, and I think SCC will get the message very soon. This is my first time as a tree protester and a very small effort by myself might have quite a large impact on something. I'm not scared of being arrested, and anyway they've got to catch me first! It's very comfy up here and I'm having fun. I've got a pillow, I've got food, and they're passing me mint tea in a bucket.

Calvin: The reporters loved the 'man up the tree' angle of course. It was a great piece of NVDA which gave lots of publicity to the tree and the overall issue. I think it was the first time in the campaign someone had been up a tree, but not the last! We never saw the guy before or since, but he earned his place in the Sheffield tree protest story that day – top man.

Sally: We had a few plans for that day, lots of different things. We composed a letter to the arbs that we'd written on a big leaf! I knew it wouldn't make a difference, but it was all about making people examine their consciences.

Margaret: We planned to slowly walk across the road to stop the trucks getting in, and to hold hands around the tree.

Ann: No chipper, no barriers, no crews have ever turned up to try to fell Vernon. We'll never know whether Amey really intended to come. Was it the man up the tree that stopped them? We'll probably never know, but one way or another, we all did what we needed to do. I think that day was a boost to the campaign as a whole. A well-publicised, peaceful victory.

We don't know what would have happened had the crews arrived that day. I think there were enough people who were prepared to go to their limit, to their ultimate, that day.

2017-2018

'I think the Council and Amey thought we were a soft touch compared to other parts of the city. They had us down as a bunch of retired women they could bully.'

Ann Anderson, Protester

Chatsworth Road is a residential street in Dore, adjoining Vernon Road.

Sue Unwin: From where I live on Chatsworth Road I can see Vernon, although when the council survey went out I was only allowed an opinion on my road, even though those trees are further away from me. When I'd first heard about the plans for Chatsworth Road I didn't know about Vernon. The week after we had all turned out to protect Vernon they came gunning for Chatsworth Road. They were annoyed after Vernon, I'm sure.

Ann Anderson: They stretched us by booking work on Abbeydale Park Rise and Chatsworth Road for the same week. It was around that time that relations between us and the arbs, which had been cordial until then, got worse.

Margaret Peart: One of the arb supervisors, Dave Smith, said to me, 'I'm not talking to you, and I've told my crew not to talk to any of you either'.

Sally Goldsmith: My biggest guilt in the whole thing was being talked out from under a tree on Chatsworth Road. I was tricked into moving.

Ann: It was one of the supervisors, Jamie, who got Sally out from under the tree that day. He said, 'That's it, we're going' and turned to go back to their vehicles whilst other arbs got back to the work zone. It was tense between us and there were a lot of police present by that stage too.

Alice Fairhall: I didn't go there often; they had a stalwart group of early morning patrollers and lookouts who rarely asked for help. One day sticks in my memory. Following an alert, I arrived to see a huge number of police vans lining the surrounding roads. This was a fairly new thing, and just the day before, Kaarina [Hollo] had filmed herself walking down the road counting the number of police officers present; there were twenty-two on that day. I arrived and saw a few protesters, maybe six in total, mainly middle-aged women and residents; one lady was crying. The street was filled with police, security guards and arborists – almost exclusively men in yellow hi-vis. The police were lining the road on either side of the tree being felled. I approached the tree, and could feel the anger rising in me at the ridiculous waste of police resources – fifteen plus officers facilitating one tree being felled in quiet suburban S17, when crimes were happening all over the city! I just kept thinking, 'Why are they here for six quietly watching residents? What should they be doing instead?'. I asked to speak to the Inspector in charge of the operation. I was fobbed off and told to ring 101 if I thought there was a crime, so I went up to the police line on one side of the tree and went along asking them individually and rantily, 'Why are you here? Where have you come from? What should you be doing?'. Understandably, most of them were ignoring me and looking away – what else could they do? They had been sent there and were, after all, just doing their job. Then one officer engaged with me, I guess, trying to placate me or maybe just to pass the time, and he told me he usually worked in CSE [Child Sexual Exploitation]. I felt something inside me flip and my normal smiley disposition disappeared. I was enraged. 'What?!'. He said, 'It doesn't matter what I normally do', but at that moment, I couldn't agree. 'This is outrageous. Why on earth are you here instead of protecting children? Go and do your job'. I said, 'I don't need policing', (although given how angry I was I might have done!). I shouted, 'Does it need all of you to deal with one annoying teacher? Really? You can at least arrest me'. I was furious and nonsensical. The months of repeated confrontation, the frustration at the lack of proportionality; I was emotionally frazzled. After a few minutes, I was shepherded off the street for a coffee and to calm down, but it'll stick in my mind for a long time – the lengths the Council will go to 'beat the residents'.

Helen Damnation: After witnessing the felling of a beautiful, mature, healthy tree, I stood alone in front of a van on Chatsworth Road whilst the police kettled everyone else off onto the pavement. The police surrounded me, asking me questions which I could not hear above the noise of the chainsaws, so I stood my ground. I'd broken my arm two months before and had just had the pot removed. Despite this, the police grabbed my bad arm, which was still in a sling to protect it, and forcibly moved me away from the vehicle and towards their van. I had been playing a pink sparkly recorder, in support of Andrea [the tooter] who had been arrested the day before on Rivelin Valley Road. I was playing it rather badly I'm afraid as the thumb hole was not quite covered, producing a horrible squeaky sound. Anyway, they took me to Shepcote Lane detention suite where I fell asleep. They interrupted my sleep to take me to a brightly-lit room and when I asked what this was all about, they said they were just going to take my fingerprints. I walked myself back to my cell muttering that it was a contravention of my civil liberties and a little later they woke me and gave me food. Six hours later, I agreed to be interviewed. I never gave them my name or address and they released me when they accepted Paul Brooke's address as proof of being able to contact me (after talking to Paul on my phone). I remain anonymous despite Rob [Pearson] being heavily penalised for not giving his details.

Tina Sampson: On 22nd March 2018 I was arrested alongside another protester whilst protecting the trees on Chatsworth Road. The other protester was playing a pink sparkly recorder and I was accompanying along with my shiny 'light up' tambourine. I went to talk to protesters who were protecting trees further up the road when I saw the pink recorder person being led away by two burly policemen. My immediate thoughts were, 'Oh no!'. I lent the pink recorder person my coat as they were getting very cold, unfortunately in the pockets of said coat were my mobile phone and house keys! I shouted to the officers to try to get my coat back, but instead of that I was pounced upon, my arm twisted, and also arrested. I was in a great deal of pain as I suffer from a frozen shoulder. I repeatedly told them they were hurting me, but they didn't stop and bundled me into the police van. At the police station the desk sergeant asked for my name and title and seemed rather surprised when I told them I was the newly ordained Reverend Tina Sampson-Smith. Suddenly, I was de-arrested and told that I could leave. I had no British money as I had recently returned from missionary work in Zimbabwe, and only had Zimbabwean money in my pocket. The desk sergeant insisted on the arresting officer returning me home, but due to a lack of transport available I was returned in the back of the police van I had been in previously. When I arrived home, the arresting officer had to let me out of the back of the van. I jumped out and gave him a big hug and he shuffled off back to his van with a very red face. I have heard that from that day to this he has never lived it down, and is still made fun of by his colleagues about the day he arrested a vicar!

Ann: I think the Council and Amey thought we were a soft touch compared to other parts of the city. They had us down as a bunch of retired women they could bully. They put a lot of resources into trying to break us down. We had a good local group, but really appreciated the support from protesters across the city. When Justin [Buxton] or Paul [Brooke] came up the street I would be so grateful and relieved!

'Busy Bee', Meersbrook Park Road.
Photograph: Luis Arroyo

Tree-felling.
Photograph: Luis Arroyo

Marden Road group (pre-arrest). Photograph: Sally Weston

Christ Rust, Raven Road (Baby Blue Puffer Jacket, one size too big).
Photograph: Chris Rust

The Chippinghouse Road Seven

Woodstock Road group. Photograph: Christopher Carter

Photograph: Luis Arroyo

The Kenwood Road 'Brash Incident'. Photograph: Lynne Chapman

Gandalf on Rivelin Valley Road. Photograph: Luis Arroyo

Photograph: Luis Arroyo

Gandalf on Rivelin Valley Road. Photograph: Luis Arroyo

Vernon Road climber. Photograph: Luis Arroyo

Chatsworth Road Heras. Photograph: Luis Arroyo

Abbeydale Park Rise, hi-vis line.

Coverdale Road hi-vis army. Photograph: Calvin Payne

Thornsett Road in the snow. Photograph: Calvin Payne

Meersbrook Park Road.

Meersbrook Park Road.
Photograph: Luis Arroyo

The Thornsett Road (short lived) victory group.

Meersbrook Park Road 'Shamey Cowboy'.

The 'Pink Panthers'.

2017.

'Don't be silly – save Milly.'

Eddie, aged 4, Resident

Millhouses Lane is a long road linking two main roads that are air-pollution blackspots. A condemned, healthy ash tree, 'Milly' became a well-known focus for the local campaign.

Alice Fairhall: The first I knew of the tree issue was when the yellow ribbons appeared on trees in my road and I remember thinking, 'That's not on'. But after Rustlings Road, I made a conscious decision, 'I'm not standing for this'. I was determined to stand with the protesters. Milly was a large ash tree, one of a pair separated by the road, the other safely inside private grounds. I couldn't believe that it was listed for felling. It was on a wide section of tarmac and half standing on a large grass verge well away from the road. The pavement was slightly humped but anyone could see that that was from years of chucking more tarmac over the roots.

Polly Morley: The first time I saw the evidence gatherers was when we'd seen a chipper in the area. We went to see what was happening and they took photos of a couple of us with our kids. Until then I'd been careful not to identify arbs and Amey staff in photos but in response to them doing this I took their pictures. They were doing some strange things that day, letting us inside the work zone to look closely at a tree for example, and saying the chipper was broken. It seemed like a set-up in hindsight.

Nicky Campbell: We challenged a female evidence gatherer and followed her, taking pictures of her filming us. She was furious at us for recording her recording us!

Alice: Not long after I got involved crews started appearing and initially targeting our smaller trees in Millhouses, which they could get down quickly. I was already on call-out groups monitoring felling crews and we would turn up to any sightings of arborists and their chippers in my area and try to stop them by standing under the tree. Our aim was to hold them up so they weren't felling at all, or at least move them on. There were scores of 'smash and grab' fellings, where they'd recklessly fell the whole tree in one go, which were hard for us to keep up with. We would usually arrive too late for at least one tree, but then stand under a nearby one and usually they would then pack up and leave the road. One day, I got to a small cherry tree on a cul-de-sac just as an arborist started up his chainsaw; no barriers were up, and his team's van was parked slightly further along the road. I quickly drew up next to the tree, tightly to the kerb so my own van was fully under the canopy, and I jumped out and started videoing. I was asking about the lack of a safety zone and anything else I could think of. By now, the arborist had cut a wedge into the tree trunk and told me, 'You've got to move that van. We need to complete the job now, it's unsafe'. I said, 'You left it like that. You are going to have to deal with it', and I locked the van and walked off with the keys!

Caroline Millman: My family were often without a car in those days! 'Sorry kids, we're walking' or we had to go and collect a car from a strategic parking spot under a tree.

Phil Yates: I seemed to spend a lot of time pumping up car tyres in those days. A lot of cars mysteriously suffered flat tyres while parked under threatened trees!

Caroline: One morning I decided to work from home in the car so I could keep an eye on Milly when suddenly out of the corner of my eye I saw an Acorn van. My cup of tea went everywhere – which would have taken some explaining to my employer. I got out of the car and into the way before the crew could even pull up. The atmosphere had changed by that winter. The arbs stopped being friendly, really suddenly. One day they stopped us passing gloves and food to protesters inside the work area or trapped up against the fences 'geckoing'. Then a supporter brought her son who was a basketball player and he threw a bag of chips over the fence to the gecko!

Phil: They ended up getting permission from the house next to the tree to have SIA security guards inside the garden in order to try and prevent us getting to the work zone.

Alice: I didn't actually start protecting Milly until it got a bit heated. I spent eight hours on a cold day 'geckoing' by the tree, wedged in between the Heras fences and the garden wall. Much to the residents' disgust, we'd got in by running through their garden, opening the gate and reaching the position we wanted. Eventually the SIAs would block the route and make it harder for us, until they started coming at 5am, in winter, in the pitch dark, and worked on the tree by the light of head torches. Each time our tactics beat their Heras fences and security guards, they came back with a new idea, even earlier in the morning with higher barriers and more security. This particular morning our route was blocked off so Mark and I decided to try to distract the SIAs and aim for a weak point at a corner of the barriers. With much help, and getting squashed and bruised in the process, and being accused of assault and criminal damage, I managed to get through the gap. Two hours of rain and cold later, at first light, the police arrived. A female officer asked me if I was a protester, I said I was, and she told me there had been an allegation of assault made against me. She looked at the SIA guy, six feet plus, looked at me, five feet one, and said it was ridiculous. I told her that I was the one who had been hit and bruised. She asked me if

I was going to move away. I said, 'No, I really don't want to. I've had to work hard to get here and this is going to stop them felling it'. She said 'OK, I'm leaving you to it', and walked off.

Susan Ashworth: Alice was in position from about 5am to 2pm, alone. I watched her get across the hedge, slide in, and tell the others that they could go and she'd stay as long as it took. She was just amazing.

Polly: They were just shoving the fence into her, I believe it was an assault.

Nicky: I got involved with guarding Milly fairly late on. I'd been dedicating one or two days a week to tree-guarding duties. Not much compared to our car, which was on 24/7 sentry duty on Brincliffe Gardens! By this time Nether Edge was pretty well covered with lookouts, whilst Milly was short on numbers, who had repelled attack after attack and, despite their tenacity, were getting worn down. Most of my time with Milly seemed to involve getting there at five in the morning and getting incredibly cold with nothing much happening. Occasionally I got to stand in support with someone else 'geckoing'. If you have never been there it's easy to underestimate how very important this is, knowing there are people there for you and supporting you.

Maggie France: One particular day was a bitterly cold one. We were all sitting in chairs wrapped in blankets watching out, when they arrived and we suddenly had to rush across the road to the tree!

Polly: We nearly got hit by the trucks. It was stupid really, but by then we'd come so far and panicked that any delay by us could have lost the tree.

Nicky: Eventually my turn arrived. When I turned up at Milly that morning, Heras barriers had already gone up, security guards were patrolling the perimeter and filming everything. Caroline [Millman] had been on lookout duty and managed to stay in position with the barriers running in front of her, leaving her in a little pocket hemmed in between the wall and the fence, but not close enough to the tree to stop it from being felled. I decided to try to join her and hopefully slink a bit closer to the tree behind the barrier. The corner of the fence and the only way to reach her was guarded by an SIA. And joyously, they had left a piece of small plastic fencing at the corner too. Distracting the security guard was easy. I went to a close-by section of the barriers and with the help of another protester standing in front I made it appear that I was going to interfere with the barriers in some way. The SIA came over and I zipped back to where he had been standing and was over the Heras fence in a tick with the aid of the little plastic fence for a leg-up, and over the head of the rather surprised returning security guard who didn't have time to stop me. And so I was in with Caroline. She looked tired and worn. She had done her bit and kept the place near the tree, and she was now relieved to be able to leave. I was behind the fence in a little space where they had built the barriers around Caroline. They were starting to chop down Milly. The Heras fencing was tight up against a very tall hedge, with extra security guards trying to stop me getting any further. It was a privet hedge and not prickly. I leapt up as far up the Heras as I could and 'ran' high in the hedge behind the Heras, squeezing, almost swimming, through the gap between the hedge and barriers and eventually dropped down just behind Milly between the hedge and the fence. And there I stayed. I had a book to read for my book club, so I read that. I was there for hours getting colder and colder. They started felling the tree over my head. Bits of twigs and sawdust were raining down on me, the sound of chainsaws grating and loud. It's a ghastly noise when something you care for is being destroyed in front of you. A security guard was 'guarding' me just inside the fencing. Guarding in this sense consisted of repeatedly trying to crush me between the fence and the bush by slamming the fencing into me. Luckily the hedge was non-prickly and a bit squishy. It went on and on. Hours and hours with bits of tree dropping around me. Freezing cold. But Milly's trunk divides in two, and half of the tree is right over my head. Half of Milly can't be cut down for as long as I stay put. Eventually they start to pack up. There isn't much daylight left and they can't finish the job because of me. As they finish, I move back to be with the other protesters for much needed human companionship and support. I'd lent my scissors to another protester earlier, and it was good they had stuck around and gave them back to me. But I'm exhausted and hungry and cold. Then the barriers started to go back up. Had they planned it or just seen an opportunity? I'm not sure I was doing well enough mentally to out-think them at that point so who knows? I could see all my hard work slipping away as the Heras barriers were put back into position with me on the wrong side of them. Another protester pushed a barrier open for me whilst security weren't looking, and before they had managed to secure them with cable ties. I ran as fast as I could back under the tree before the Heras fence was back in position, dodging the security staff running towards me. And I made it! So I was back under the tree, and there I stayed – until the police arrived. As the barriers came down again, I was questioned by the police. They were happy enough that I hadn't broken the injunction, as I'd pointed out to them that the barriers were not complete when I'd crossed the work zone, but they were very interested in the scissors which I still had with me rather unfortunately. If I had known a day would come when I would be sat in Snig Hill police station with my solicitor discussing with the police whether running with scissors was classed as knife crime or just something your mother would highly recommend against, and whether anyone had or had not criminally damaged a 20p cable tie. I always think a nice vintage-style dress helps on these occasions.

Alice: I was there when they finally cut Milly down. I still think of that day every time I drive past. Even when the tree was reduced to a stump, I was so frustrated that I refused to let the stump grinder into the work zone. I marched up and down the work zone, on my own, in the freezing cold, just to be bull-headed, thinking, 'No, you're not going in. I'll still hold you up'.

Polly: I had to try and get a piece of Milly as we wanted to find out her age. An arb begrudgingly gave me a piece, but only after an argument. Then a bunny arrived, very sadly just too late. They only stopped work when a class of school kids went past. They had a cup of tea then resumed as soon as the kids were out of sight.

Roger Doonan: They'd been nibbling away at Milly for a while, and I think we knew the raid was coming. The road was on my regular early morning drive round. I did my usual Crookes, Rivelin Valley Road, Milly, Nether Edge circuit. That morning at Milly, the crews were there and the Heras barriers were up. It was a full-on assault, tragic. There were locals there, campaigners, lots of people. The week before, I'd been there when geckos had seen off the felling crews, but now a bunny had got there too late; the crews were too quick. As the bunny attempted to get into the zone the SIAs tried to stop them, and I reacted and helped do just enough to let the bunny get in. I thought it was going to get physical; I was trying to protect the bunny from harm as much as anything. The security guards said to me, 'We know who you are. You're breaking the injunction'. It was a small thing but still scary, the only time I'd really been threatened in that way, but nothing compared to others.

Caroline: My daughter was crying and my son, Eddie, shouted at the felling crew, 'Don't be silly – save Milly' at Dave Smith, the arb supervisor.

Alice: We couldn't hold them off forever, they got small bits each time, but I was so proud of the solid resistance we all put up.

Nicky: Milly was felled eventually, when they got her other half, but it wasn't all for nothing. If we had just sat back and let them, then how many other trees would have been chopped down in the many weeks and months it took them to eventually fell Milly? And when I tell the kids not to run with scissors, I really, really mean it.

2017-2018.

'I just didn't want to trample their daffodils.'

Nicola Gilbert, Protester

Abbeydale Park Rise is a tree-lined residential street in Sheffield, adjacent to an air-pollution blackspot. It is home to a noted avenue of cherry trees famous for their blossom display each spring and a Christmas light display which raises money for the local children's hospice.

Stephanie Barringer: One day in June 2017 it was my day off work and I heard all this commotion on the road, and I came outside. I saw the trucks and I saw Ann [Anderson] from the local campaign group, so I went running down the road and they were coming for one of the trees further down. Ann said, 'They're trying to take the trees!'. I sort of asked what we could do, as it was the first time I'd experienced this and she said, 'Well we can stand under the tree', and then we saw that the Amey staff were giving out pieces of paper.

Calvin Payne: They were just giving them out to anyone who was paying any interest in what was happening.

Marc Machin-Cowen: My neighbour took one without knowing what they were.

Stephanie: I was standing with a campaigner who said to me, 'Under no circumstances accept that piece of paper', because I didn't know what it was. They made it look as though they were handing out information about the trees or the work.

Calvin: I think we had only seen these letters, 'Pre-action protocol letters', for the first time a few days before.

Stephanie: Some residents got one. My friend Kathryn was absolutely beside herself and got really upset. Jeremy Willis from Amey was there at the time and I spoke to him and said, 'You need to take these back, this is absolutely ridiculous'. They were filming people taking them as well and taking photos. I kept saying to them, 'Why are you taking photos?'. We hadn't even done anything. I wasn't standing inside the barriers. I wasn't standing anywhere they wanted to work.

Calvin: We know now that there was a list of named people that they were looking for but we didn't get the 'Persons Unknown' element at the time, I don't think.

Stephanie: I said to one of the Amey staff, 'What are you doing?' and he said, 'I've just been told to hand these out'. I said, 'What are they?' and he went, 'I don't know what they are. I've just been told to hand them out'. Ann was talking to Jeremy Willis, and going absolutely mad at him. Luckily I was standing next to a campaign member who said, 'Don't take it. Whatever you do, don't take it'. I would have taken one if he hadn't been standing there with me.

Stephanie: Then they tried to barrier up around a tree and we stood underneath it so they couldn't barrier it up.

Simon Crump: You still could then!

Stephanie: So we all stood around the tree and then we went running down the road to another one, and stood under that tree and refused to move, and said, 'No, we're not moving' and they kept saying 'Will you move out of the way?'.

Marc Machin-Cowen: First attempt to fell trees on the road and I was at work!

Stephanie: Basically most of the residents on the road came out. Lucy came out and Emma was out, loads of people, basically the whole road. Gordon came out, and there were people sitting on their deck chairs under the trees. People got chairs and sat under every tree. And my daughter came home from school and called a few of her friends and they came up and we all just sat under trees and refused to move. So the felling crews went, and that was that for the day.

Briony Hancox: I'd been involved since going to Council meetings in 2015 and it had taken all that time for something to happen on our road. The second time they came to the road Marc was at my house because, believe it or not, Marc and I both used to live at my house during the week and stay at Marc's at the weekend. But now it's completely the other way round, to protect the trees. So early one morning, we were arriving on Abbeydale Park Rise and saw the felling crews. There was nobody out, so I drove up and down. I dropped Marc off to get under a tree and I drove up and down pissing everybody off by tooting the horn to raise the alarm!

Stephanie: It worked! I looked out and I thought 'What's going on?'. It was on the local Facebook page! Everybody was asking 'Who was the idiot driving up and down Abbeydale Park Rise?!'. We all blamed Amey and said it was them that were driving up and down tooting the horn because they couldn't get up the road for the parked cars!

Briony: We did blame them. It worked! Because everyone came out. So that twat who said it was quarter past six, he knows exactly when it was, he said quarter past six and therefore illegal. But we couldn't really say, 'No, it was quarter past seven' could we?

Stephanie: I put it on the Facebook page. I said, 'No, no, it was them because it's bad for parking on here anyway, they couldn't get through with the trucks so they were tooting their horns'. So we blamed it all on them, didn't we?

Briony: Well until then nothing happened for so long, so I thought, 'Oh, it's not going to happen'.

Calvin: You get that all the time; people would think something had been sorted because it had been months, even years, without the trees being felled.

Marc: They'd put these notices up, but nobody was going to come...

Jane Sharpe: When you were tooting I had forgotten all about it, forgotten everything to do with it.

Briony: Then they got all Jane's lights out of the tree and chucked them in her garden. Then they described them in a statement as 'unauthorised attachments'! The road has always had lights in the trees at Christmas and some stay there all year.

Calvin: Amey do love a PR disaster, don't they?

Stephanie: So what they did after that is they asked my neighbour to take her lights down and they said, 'You've got half an hour to take your lights down out of the tree'. Her lights in the tree are plugged into my house. So they have to ask me to take the lights down. Unfortunately I was at work and so I was unavailable.

Jane: My argument was why they couldn't share the respect that we had when putting them up. It was complicated and carefully done. My kids put them up. They could have unwound them and put them politely on the hedge; instead they hacked at them. So I wanted to know why they did that. That was my argument with them.

Marc: Then one day when Dom the arb was on the road, he said, 'I will pay for them myself', didn't he?

Stephanie: He said, 'I'll give you a quid for some more lights', didn't he?

Jane: I wasn't taking it seriously then. For me, it was like a rude awakening. So I just thought, 'They're never going to take my tree'. I didn't go on any of the walks or events, I just said it's not going to happen. Then I joined the WhatsApp group but I came off it because there were too many notifications. So I just thought, 'No, they're just not going to take that tree'. Then all of a sudden one morning, it happened. That's when I went into shock. I came running out.

Simon: I was trying to calm Jane down; in fact I had been assigned that job for the day! She was ready to kill and/or go to prison, she said.

Jane: Did I? I couldn't stop swearing, I know that!

Briony: It was 15th November 2017, I've got film and photos on my phone.

Jane: They were filming it and I was kicking over the barriers. I didn't realise they were filming me. I was swearing and kicking over the barriers. And when I'm standing on that stepladder, I look fat! I froze it on the film! I said, 'That's it, I'm on a diet! I don't care if I look hysterical, as long as I don't look fat!'. That was when I got involved in the campaign.

Briony: Soon after that was the early morning when they came back for the same tree again.

Marc: Somebody phoned me up at 4:30 in the morning and said 'They're coming for my trees!'. I just shouted 'What?!'.

Jane: I remember putting 'HELP HELP HELP' on the WhatsApp group. I was standing there but the police were there too, and I was going 'HELP!' down the phone and they were just standing there looking at me.

Simon: I brought a 'bunny' up to the road that day. The day before, the arbs had been winding me up saying they were going to call the fashion police on me. So I turned up in a proper three-piece tweed suit, and it actually distracted them. As I brought the bunny to the road in the car, she changed into her outfit (she had a choice of two) and she asked, 'Do you recognise me?'. And I said, 'Of course I fucking recognise you!' and then I distracted the arbs, this time by just winding them up, really. The bunny went over the barriers, but as she did, she steadied herself on an arb and they took her to the police station for questioning.

Briony: That was when they dragged the bunny across the road.

Stephanie: They set up barriers further up the road and they pushed the barriers up against me, but then they decided not to go after that tree because there were telephone wires going through it. They decided to move and my car was under the next tree, because we never moved any of our cars from under the trees. We never once moved them off the street.

Jane: One of the bunnies got into the work zone through my garden. There was a gap in the fencing and he or she got through. And they just stood there because the arbs had already climbed up the tree they were starting work. It was a desperate situation and they came and just ran in through the gap and went in.

Briony: The second time we got the bunnies in, they dragged one of them out, didn't they?

Marc: That was when my daughter was working from home and shouted, 'This is not reasonable force! What are you doing?'. She was shocked.

Jane: My daughter had come out of school early, as she had a dental appointment, so she saw what happened and was standing with her friends and she said, 'I've never seen anything like that in my life, Mum'. She said it was a proper assault, that they pulled her top up and everything.

Stephanie: This was when the security guards were covering up their IDs.

Jane: They grabbed her by the scruff of her neck and dragged her across the ground! Then their supervisor turned up. I think it was Jamie the supervisor that wore the orange jacket: The Amey guy. He called an end to it. He said, 'That's enough. You wouldn't want this to happen to your sister or your mum'. And I overheard one of the security guards say to him, 'You got us in to do a job, let us do it'.

Stephanie: That day also, they were refusing to let me out of the top of the road. I said, 'You can't refuse to let me out of the road'. When I came back in they tried to stop me again and I said, 'I live on the road you've got to let me back in'. As I was coming down, they stopped me again and I said, 'I live just down there' and he said, in front of one of the builders that had been working on the road, 'You won't believe what they're doing down there. They're throwing petrol bombs'. I said, 'What do you mean, the old lady down there with the walking stick, a flask of coffee and a woolly hat?' I said, 'I'm pretty sure if they're throwing petrol bombs, it would have been all over the national news'. As it was when they weren't showing their badges I couldn't get his number or anything.

Jane: The thing that I found really awful was the fact that the police were there seeing and hearing all of this. Amey had called the police and they stood there and they just watched everything.

Tina Sampson: I was there when Justin [Buxton] was arrested. He was being perfectly calm, asking questions. The police were arrogant, they didn't answer his questions. The fact that they used excessive force, when Justin said he would go willingly with them, was shocking. There was no need to use handcuffs. This was an example of police overreaction to peaceful protest. It's shocking that they felt the need to do that. It was like Orgreave.

Briony: I think they came to the road three times in that last week of November. There was a night raid in December. We'd already got the lights up in the trees. They said they were going to get the trees before Christmas.

Marc: We had a lookout rota from very early in the morning by then.

Calvin: There was a night raid in December that was on the same day that Alastair [Wright] was pulled under the injunction for what he'd done. I'm in the video of Alastair making his stand in Millhouses, which is on YouTube. I was really stirred up, because they'd done the night raid on the children's hospice fundraiser trees earlier and I was really laying into the arbs, in the heat of it all. Later, I posted my thoughts on the STAG Facebook page and was banned for a few days. I'd called the arbs who sneaked in and cut down trees for a hospice fundraiser 'scum' and I don't regret it.

Stephanie: We ended up getting much more publicity for the fundraiser that we'd been doing for years. Bryan Lodge [then Council cabinet member responsible for the policy] accused us of only raising funds to attract publicity for the tree-felling issue. That week I had a knock on the door from somebody from Radio Hallam, a researcher who said, 'I've heard all about your street with the lights and everybody said, oh, we should come on here. Can we come and switch your lights on?'. And I thought, 'Result!'. Didn't mention anything about trees, I just said, 'Right, that's fine'. And he asked me, 'Why are all these yellow ribbons on your street?'. I told him. I was honest with him, and said it is because they tried to chop the trees down. Later they phoned and interviewed me about the felling plans as well as the fundraiser, and all thanks to Bryan Lodge!

Jane: On the early morning raid people ask me whether the crews woke me up. And I lied to the press to say they woke me up but actually I was already up, because I'm a very, very early riser, and watching out, so I was awake and I'd made a cup of tea. I sat down and saw this head moving in the tree and I thought, 'Oh no!'. So I had to make out to the reporter that they woke me up because they'd think I'm a weirdo if I'm up at four in the morning!

Stephanie: They got two trees that day. They got Maisie's tree, the tree that Maisie's grandma used to stand under and wave to Maisie through the window, and she used to put things round the tree for her grandma and leave little presents on the tree. She had posters all over the tree. When they came for Maisie's tree, we waved these posters in the arb's face, saying, 'Are you not at all ashamed by what you are doing?'.

Marc: He said, 'I don't give a shit'. That was Dom wasn't it?

Calvin: Right little charmer, isn't he?

***

Briony: The next visit from the felling crews and the police was near the end of the story, 5th March 2018. The day with Nicola and Benoit. That day started with riot shields on the road.

Nicola Gilbert: I'd been attending protests in different parts of the city for a few months. I'd gone to Nether Edge, Burngreave, Meersbrook, Firth Park and more. I live just up the road from Rustlings Road, and like lots of people I was shocked by what happened there. It was in Nether Edge where I first saw all the hi-vis jackets, felt the tension and saw rows of police vans. It looked like they expected a riot when we were just peaceful people who opposed what was happening. It all had the effect of upping the ante. It was scary – what the hell was happening and why? It wasn't a load of criminals, just responsible citizens carrying out their rights in protecting the environment.

Calvin: There must have been people who turned the corner and wondered just what was happening. Serious crimes and incidents don't have as many police present as we sometimes had.

Nicola: My son had been attacked with Heras fencing at a protest on Thornsett Road and when these things happen it doesn't have the effect that the police and authorities hope – it often has the opposite effect. It made me leap up and get out when the call-outs happened each morning.

Calvin: Every time the Council, Amey or the police increased either the felling crews, the arrests, any sort of crackdown, it backfired. And yet they all kept trying it!

Benoit Compin: I knew this was an important street. It was mad seeing all those police there on Facebook I thought, 'It's crazy'. At the time I was walking everywhere, carrying my guitar to protests, so it takes me a while to get where I'm going. That day I thought I could be arrested, just seeing what was happening. I told myself to be calm, take a breath, and decide what to do for the best. I got prepared. I took the guitar and the didgeridoo. I managed to get a last-minute lift over. I was dropped off at the bottom of the road and walked up.

Nicola: The day I got caught up in events on Abbeydale Park Rise, 5th March 2018, was supposed to be a normal 'at home' day for me, just catching up on work. Then there was a message from another regular at the protests, Judy, who asked our messenger group 'I'm going to APR – can I pick anyone up on the way?'. So I went.

Calvin: The campaign call-out networks, the alerts on social media, the word of mouth that something was happening, and where, was amazing sometimes. All it took was a sympathetic resident, even a passer-by, to tip us off, and we were up and running.

Nicola: Soon after we arrived, the police were kettling us and moving us up the road. It was quite extraordinary. As we slowly moved, I was somehow tripped up and fell over. All I could see were legs until somebody pulled me back to my feet. It was really quite frightening; it was an experience I've never had before. I had the fear that I could have been crushed at any second.

Neil Furmidge: I remember the kettling on Abbeydale Park Rise very well. I recall Nicola stumbling to the floor. I stood next to her and helped her back to her feet. She wasn't keen at first, as I was wearing a hi-vis jacket and it wouldn't have been clear from down there who I was.

Benoit: As we all walked slowly up the road I started falling over, thinking it would be funny and disruptive. I got up slowly, the police went to help me, but I told them, 'You're not here for my safety'. Then other people started falling over or getting slower and slower.

Calvin: I was on the pavement alongside and it was clear that something new was happening here. I don't remember anything similar happening during our protests before. This is one of the protests that was filmed throughout. David Glass on YouTube has hours of footage showing what happened.

Neil: I specifically remember the line of police keeping on pushing whilst Nicola was still on the floor, and I put the flat of my hand up to one officer and shouted that there was a lady on the floor. The officer did stop for a short while and explained that he was being pushed on by his superior officer. I believe that senior officers present were forcing other police to behave in a way that I don't think was reasonable or at all necessary.

Nicola: I was ready for anything – it was almost a transcendental experience, like being in another world, not normal at all. Until all this started I had a good attitude to the police, I thought they were there to help. This was a change to everything I had thought before.

Neil: At the time I was very concerned for people's safety, especially Nicola. It could easily have led to a serious incident and was entirely unnecessary, as the group were only slow-walking a lorry up the street.

Nicola: Some of the Council evidence gatherers had an evil look about them. You could tell they were quite enjoying the aggro.

Calvin: There was an interesting exchange about garden permissions between campaigners and police as this was happening. There was some controversy as a long-standing permission had been retracted after police spoke to the residents. The discussion that followed is on David Glass's YouTube channel.

Police Officer Unknown (to protesters): 'Do you know every person on this road? And got written permission from them? Permission from that specific address that they're happy to have the best part of twenty people on their property tying themselves to a tree? Really?'.

Ann Anderson (to the officer): 'I'll tell you exactly the situation. We've been knocking on doors and spoken to every home on the street. We have permission for most of them – we're not some clandestine group. We've been going a long time. I'm telling you categorically that I would not condone anybody standing in a garden without permission.'

Calvin: They didn't believe us. They just didn't get it at all, I don't think. They definitely changed the minds of those residents, but I don't know how.

Nicola: It took them a while to set up the work zone around the tree, and a few of us were wandering about in there. I kept walking round and round; I think it was the tension. I was talking to the police officer who seemed to be in charge, I told him, 'You shouldn't be here doing this. This is not what police should be doing. You are prolonging the situation and making it worse. If the police stayed out of it then the council and the campaigners might be able to start talking'. They closed up the gaps in the barriers and started giving me the usual warnings about the injunction. They told me twice that I was in breach and I said, 'OK, take me out'. I offered myself to be removed. I was saying that I was submitting to them to remove me. I wanted to give them a job, to make their existence worthwhile. I didn't want to just amble out by myself.

Calvin: They are on film that day, accusing people of breaching the injunction when we were standing in supporters' gardens or were inside an incomplete work zone. They either still didn't understand it or were misusing it just to scare people.

Benoit: It took them quite a while to set up and clear the work zone. I sat there playing the guitar. I waited to be the last one removed and just to see what happened. I took my time, asking lots of questions. I saw Nicola being escorted out of the zone. I watched them start to carry her over the grass verge, lifting her legs up. I knew that after removing her they would come for me, then I saw her legs failing and saw her being dragged. I wasn't having that. If there was something to be done then now was the time.

Jane: I saw Nicola give an interview through the barrier to a TV reporter. All of a sudden there's a big guy, a big SIA bouncer; it quickly got very worrying. He was aggressive to Benoit, who answered him back, because I think he was irritating them because he's not aggressive, of course, he's very artistic. He winds them up verbally in a creative way. He's saying stuff like, 'I'll take responsibility for my own safety, you don't need to worry about me' and 'I'm not moving, I don't need to move, I'm fine!', and they were getting really angry.

Nicola: They were trying to direct me over the verge when there was a path just a couple of steps to the side. I didn't want to be taken over the verge as I had heard that residents had planted daffodils there but I couldn't see exactly where because there was snow on the ground. I thought, 'These people are going to lose the tree, I don't want them to have trampled daffodils as well'. It was only a minor detour, but they wouldn't divert just that little bit when I asked, so they started dragging me, which was different from being carried out calmly and peacefully before.

Benoit: Without thinking I said, 'Put her down, put her down, you're hurting her'. They let go of Nicola, and she fell to the floor. I helped her up, looked behind me and saw the SIA supervisor and he was right on me. Nicola was on the ground and they were suddenly all on me. I didn't fight at first. I used the didgeridoo for defence, trying to trip them up with it.

Nicola: They were roughly holding me, roughing me up, and it was all over not taking me round the verge! I was happy enough to be removed. I'd made my point. I just didn't want to walk over the verge where there might be flowers. I just didn't want to trample their daffodils. It wouldn't have taken them any longer, but they doggedly insisted on taking me the way they wanted. I lifted my feet and they just kept on dragging me. Then Benoit saw me being manhandled and he, all of a sudden, arrived in the melée!

Stephanie: Benoit obviously grabbed her to try and steady her. He saw the aggressiveness of the bouncers as they are. Of course given what we'd seen from the bouncers, you're going to defend somebody.

Jane: Well at the time when they were taking her out we were all saying, 'Don't do that, don't do that'.

Nicola: I ended up on the ground. I got up but was knocked down again – on the second time I hit the ground I ended up face down in the snow. By then the attention had gone on to Benoit; they seemed to forget about me lying on the ground. They were pinning Benoit against the Heras fencing instead, in what looked like a rough, brutal way. They couldn't have cared less about me.

Benoit: I tried to reach the fence to have something to hold on to, but the SIA supervisor was twisting my fingers to stop me. I was not having that, I'm a musician and my fingers are my instrument of work. I needed to find a solution and I've got his balls in front of me. They didn't want me to use my hands or arms, so I just launched myself forward at him. I knocked him in the balls. I used the fence to bounce back and forth towards him. A police officer was telling me that I was breaching the injunction and I was telling him that I'm not. I tell him, 'Get these guys off me and we'll talk'. I managed to climb the tree and it all cooled down. I decide to stay there until they leave. I started to record a 'Happy Birthday' video to my mum and people around joined in.

Calvin: By then, Benoit had climbed the tree and all the attention was on him. I remember Nicola being on the ground for some time, wrapped in one of the foil blankets you see at the end of marathons.

Nicola: The only help I got from anybody was from other protesters. I'm not sure who brought the foil blanket, maybe the police, maybe a protester. An ambulance was called but there were none available soon enough, so I was taken by the police to the Northern General Hospital, accompanied by six police officers and Jenny Saul, who had been looking after me.

Neil: I think the blanket they wrapped Nicola in was from Amey who passed it to a police officer who gave it to me. I think it must have been in their first aid kit. I'd already given Nicola my jacket to sit on. I was worried about shock or hypothermia. The time it took to get a vehicle to drive to the hospital was poor in the circumstances. I sat next to Nicola trying to shield her from the cold wind.

Benoit: The police told me that I would be arrested when I came down. I think they didn't bring an ambulance for Nicola, because they thought they would need to remove a section of the fencing, or that I would be able to use the roof of the ambulance to get away. When I was arrested, I identified myself as 'Art Raven', my performing name, and said that I was being my poet self. They accused me with three counts of assault and one of obstruction.

Nicola: I was in hospital for hours; my family has a history of heart conditions and with my medical condition there were tests I had to have. I was at risk, in danger. Whilst at the hospital I was interviewed by two CID Officers. They asked me if I wanted to press charges against Benoit for assault. They suggested the idea, they definitely suggested it to me. It was late by then and had been a long, long day.

Benoit: I enjoyed every moment of the tree campaign. Life sends trials sometimes, and it's when you are fighting that you feel most alive. I've been living the dream in Sheffield for ten years. I walk everywhere, I've worked, had clients, met nice people, and I've enjoyed the beauty. I believe when a place has been good to you it should be paid back. I was looking for a way to do that. I did that in different ways, but I had never found a way that really inspired me before. For me it was about making a sacrifice to give something back to the city.

Nicola: The following day I was being interviewed for local television when the police turned up unannounced and again asked me questions about Benoit's role. Again they suggested that I had the option of pressing charges against him.

Stephanie: My daughter, who was eighteen at the time, became involved after she saw the police and a protester getting marched along the road. That was her first year voting and she was completely supportive of the Young Labour movement and their support of Jeremy Corbyn, but because they refused to get involved and wouldn't do anything about it, she's completely off Labour now.

Nicola: Looking back, I do hope that what happened that day contributed to them [Amey and the Council] realising that this could not go on.

Benoit: I have been upset about destruction in other parts of the world, but when it happens on your own streets, outside your home, to not do anything about it, to just be upset on Facebook saying, 'This is bad', is not enough. Ask yourself, 'What are you going to do? Are you going to sit there and watch as if it's happening somewhere else, far away?'. No! You have to be there and you have to do your bit. You have to. It's been a whole journey for me that started with trees.

2017-2018

'Coverdale Road was important, I think, because it seemed very much unloved.'

Neil Meadows, Protester

Coverdale Road is in the Carterknowle area of the city and close to a main road which is an air-pollution blackspot.

Alice Fairhall: One of my favourite campaign memories was on Coverdale Road. I happened to be in a supermarket car park very close by when a call-out message went out. Within two minutes of the alert I was inside the barriers and the work was stopped. I messaged the group and Graham [Turnbull] said, 'What?! That's about ninety seconds!'. It was fun, easy, and I was really pleased, chatting to the arbs, relaxed. After about ten minutes Phil [Yates] sprinted up the road; he hadn't seen me, and like Colin Jackson, hurdled the barriers to stop the work, or so he thought. Then he realised I was already there waiting: 'Sorry Phil, I'm already here', and we laughed at his Olympian effort to do something that was already done!

Sally Weston: It's hard to remember exactly when I first got involved on

Coverdale Road, but I was keen to go to places other people weren't going to. On Swaledale Road, for instance, not a single resident ever came out to support us. I think they were terrified, not so much of the police, but their other really aggressive neighbours on the road, who appeared to be hand in glove with Amey, so much so that you wondered if they had connections with the Council as well. Who knows, but some were really aggressive.

Alice: Swaledale Road was the next road over and a difficult one for us, lots of tree-hating (and protester-hating) residents. One morning, I responded to a call out and I arrived first to be greeted by lines of SIA security guards. I saw one of the goons clock me arriving, nudge a middle-aged man who was standing there, a resident, and he and one other man ran up to my car, which was now stuck behind the roadblock set up for the felling. One started knocking on the window and shouting through the window, 'You're a protester. You're one of them'. It was a new car and I wasn't used to it yet, so I didn't manage to lock the doors in time. The other man was at my passenger door and he got inside the car and started a shouty tirade in my face. It happened quickly but I was shocked and knew I had been ambushed. The man in the passenger seat, after a few minutes, jumped out whilst grabbing my bag off the floor which contained a body camera (why didn't I have it on?!), my wallet and bank cards and my work diary, and ran off with them down the road. I jumped out of the car and started after him, but was far too slow. I shouted at the arbs and SIAs, 'Are you all going to stand there and watch a man break into a woman's car and steal her bag?!'. They just stood there until one of them said, 'It wasn't me, love'. I shouted, 'I don't give a shit if it was you – what the fuck's wrong with you?'. I was very aware, for the first time, that I was the only woman there. I rang the police – luckily my phone was in my pocket rather than my bag. When the police arrived they went straight to the arbs first, to ask what had happened. They eventually came over to me, and I was shaking, furious, full of adrenaline, but the fear had now passed. As I was talking to the police, Jeremy Willis [Amey manager] came over to my car and handed me back my bag. I asked how he got it and who the thief was, and he said he had no idea. The police asked if anything was missing, I checked and said no, to which they replied, 'There is no complaint to be made then'. And that was clearly going to be the end of it. Their attitude and dismissiveness towards me was a shock; they even asked me derisively, 'Why did you drive onto this road?'. I thought, 'What the fuck does that have to do with anything?', but to many people that does matter. A bit later, Paul [Brooke] arrived in a big vape cloud and after a hug and supportive words said, 'You've been set up, it happens here'. It's the first time in my life that I realised that, as a protester, I was a less valid victim, my witness testimony was less reliable. For the first time in my life, I had become an undesirable neighbour. I drove to school, to work – I'm a part-time secondary school teacher – and somewhere along the way, I burst into tears. I clearly remember thinking, 'This is what it feels like for my voice to have no power, no say'. And I clearly remember thinking some people go through their life like that, and I'd never experienced it before. I'd had a good education and upbringing, I have a stake in society and some cultural capital, and it was taken from me in an instant because I was a protester. I didn't even realise that this divide was a thing. Part of me thinks that I had been getting too confident at the time; I never thought twice about arriving first or taking on lots of crews if I possibly could. I needed the warning to realise that I can be vulnerable. It's done me some good.

Calvin Payne: One morning I'd been all over the city with Simon and we'd seen off the felling crews a few times already. I then responded to a call from Swaledale Road where Alice was standing her ground. When I got there she quietly told me that she had to go to work, and could I take her place in the work zone? I went straight in, looked a line of around eight arbs in the eyes and told them all that they were really pissing me off now, and to just fuck off. I'd like to think my well-informed determination won the day, but the look in their eyes suggested I might, not for the only time, have crossed the line between protester and something a bit scarier.

***

Neil Meadows: I did a lot of days on the streets whenever I could. Coverdale Road is the main one, because I could just drop my kid off at his school around the corner. I did a lot of sitting around in the shade, in the cold, in the frost, in the rain. There was a big tree right at the bottom of the road at the corner. It was outside a house where the resident was friendly towards the campaign. I knew I could access the gardens and I knew that I could get between the gardens and the barriers. That's where I thought I could be most useful. One day on Coverdale Road post injunction, I was talking to Chris [Rust] about the injunction. Reading the big sign that was on the barriers, he said, 'That's just wrong'. It was a misinterpretation; basically it sounded like 'If you're coming here to protest then you can't do anything'. Chris pointed out, 'Well that only applies if that, this and that', and that means I could gecko.

Calvin Payne: It was so important that we had people who really did understand the finer points of the injunction, because they had been through the case itself and studied the terms, because if you don't have somebody who knows better, the other side can get away with so much.

Neil: I wouldn't have known what to do, known what I was allowed to do. I was in a garden one day and they came and started cutting the tree down anyway, and I thought to myself, 'Well, what I'm doing here isn't particularly effective'. So having that understanding was quite good, it gave me a little bit of confidence that I could be even more of a pest, I suppose, and get away with things without getting into any trouble.

Calvin: I'm still absolutely convinced that the injunction needed breaching repeatedly to expose it for what it was. It needed people to be defiant, but I think that with hindsight we could have beaten the injunction if necessary, by all doing things like that. I mean, there were ways, and some very clever. Paul [Brooke] and Chris [Rust] were two of the people who came up with inventive ideas. I absolutely trusted their interpretation of things.

Neil: Once Chris had explained the injunction and I'd had the opportunity to go away and read it, I was very confident that I wasn't in trouble even though I'd been accused of breaching the injunction for 'geckoing' before. I was happy that I hadn't. It was awful being threatened by the security, having their cameras right in the face taking pictures of you, then they'd scribble something down in their little black books and take pictures of my car. It was just a weird and horrible scenario really. Coverdale Road was important, I think, because it seemed very much unloved. There were only a couple of residents that I knew were supporters of the campaign. I justified it to myself that it was right to protest on a road where the residents wanted the trees felled by saying, 'If we can stop it, it can be looked at properly and if a tree needs felling for genuine reasons it can still come down in the end, if it's the right decision and it's done for the right reasons'.

Sally: One of the residents decided to have a go at one of the men in the group of protesters, and I remember a long period of just trying to calm him down. It was a really, really nasty atmosphere, they just wanted us off their street and all we were doing was standing talking on the corner about previous days' protests. The atmosphere was awful and you really got a sense of why people didn't want to come out. There was one supporter who was in touch with the SNET group and she used to tip us the wink. That's how we got there fairly early. It was a horrible atmosphere at times, and some of the protesters were saying it was intimidating. We also caught out someone, probably an SCC employee, sneakily talking to one of the residents about the protesters. Then we spoke to the resident and she said, 'Oh yeah, we want our trees all down'. They were trying to get in there and discredit the reason that the protesters were there. However the real unpleasantness was down the other end of the road. I'd taken a decision not to go down to that particular tree. I thought there was a really nasty piece of work who lived down there, and I wasn't going to get involved. The following day there was a call-out and I rushed over there and found that I was there first. There were at least two crews, maybe three. I got myself underneath one tree and put out another call-out and I was desperately waiting for other people to arrive. This bloke turned up, but he wouldn't gecko, he was just not prepared to do it. It was really frustrating because they'd stopped getting the barrier around my tree and they were moving on to the other ones. Eventually a bunny came along, and the two of us were slipping from one tree to another.

Neil: My most dramatic experience in the campaign was when I was on rota on Coverdale Road and they came to remove a bat box. I stood under the tree and they just backed the MEWP [see Glossary] right up to me as I was standing under the tree. The back of the truck was getting very close to me. They stuck some cones around me and said, 'You're breaking the injunction, you'd better get out'. I knew that they were going to remove the bat box. So I'm standing there thinking they're not going to get any closer than this. They backed the MEWP up to me, and I grabbed onto the platform at the back of it to try and stop them. And they just raised it anyway. So I'm going with it, I'm going up, it's getting higher and higher. It was Dom the arb operating it, little Dom. He said, 'Yeah, this goes really high, this, mate. You might want to let go soon'. I think I'd met him before on Swaledale Road when they were cutting a tree down. There were big trunks and they were just lobbing them down. They'd got most of the crown off and they were chopping the trunk. It shattered a barrier and this massive four-feet-long trunk bounced off the pavement and skidded down someone's driveway. I thought, 'That doesn't look very safe, does it?'. And I don't think he liked me from then on. So I'm kind of hanging on, and he's just stood there laughing at me. It didn't look like he was going to stop. I was probably about two metres off the ground, you know. I let go, then landed on the ground again. I could have potentially gone all the way up. I didn't know how vindictive he was going to be at that point, so I thought I'd better not go any higher! I just stayed there under it, which meant they couldn't come down, so they were stuck up there for about two hours, with the ecologist having apparently been on the lash the night before and busting for a piss, but I didn't know that until they came down. They called the police. When they arrived, I told them what had happened to me, that I was standing there first. The police were saying, 'You're blocking the road now', and they threatened me with breaking some kind of 'obstruction of the highway' law. Eventually I did leave peacefully. It was just about the time it was taking, knowing that they couldn't be going off somewhere else. I was thinking, 'The longer I can do this, the better'.

Max McBear: Our street is a small one and most of my neighbours were showing no, or little, support. It looked like it was going to be just two or three households who cared. I had already had comments from neighbours about my street-tree bin stickers and was being told to stay inside and not tie myself to any trees. It felt passive aggressive and it wasn't nice. Protester numbers would be low, I felt, as so much was going on elsewhere, but arrive they did. I appointed myself as chief tea masher and toilet provider, especially for Neil who was there every day. But no, this wasn't to be allowed. The security guys arrived and actually fenced off my entire house and all access to it, stopping any protesters standing in front of my window or using the 'facilities'. I was unable to actually leave my house without their permission. But the worst, the absolute heart-breaking worst, was the evil behaviour of the men. The constant taunting, shouting and swearing, laughing and joking, all through the day. I almost couldn't hear the chainsaw over their nasty banter. Those vile, spiteful men who thought it was hilarious when good, caring people were slammed into metal fences, it still makes me feel sick. Constant abuse aimed at peaceful folk, just because they cared. When 'my' tree was felled I managed to find a friendly arb who was kind enough to carve me a small chair from the stump, which still sits in my living room. There were few laughs that day, but when special permission was granted for my Christmas tree to be delivered, work was stopped and a cheer rang out as a lone delivery chap came up the street with my felled tree on his shoulder. My lovely road will never be the same again.

THORNSETT ROAD 7

December 2017 - March 2018

'I felt powerless, frustrated at the sound of the chainsaws. I looked at the tree in the garden next to us. I thought if I climbed it, I could jump across to the street tree, even though it was around fifty feet up.'

Willow, Protester

Thornsett Road is a residential street in Sheffield. It is lined on both sides by a variety of mature Victorian street trees. Amey targeted this street several times over a three-month period, the last being a sustained attack which lasted for a fortnight.

Simon Crump: On 5th December 2017, I was on my way to work and I just thought I'd pop over to Thornsett Road for half an hour to see what was occurring, as a felling crew had been successfully seen off the week before and we all knew that they would be back. When I got to Thornsett Road, I could see that they'd set up a big work zone around the Bridge Club tree. In the eighties I'd lived in the bottom flat of the house next door to the Bridge Club and that particular tree had memories for me. I used to meet a bloke I knew only as 'Jimmy' under it, and buy drugs from him! They'd got two layers of barriers round the tree, one plastic, one Heras, and there didn't seem to be any way to get between the boundary wall of the club and the barriers in order to gecko and prevent them from working. I was pretty sure that I remembered a back way into the Bridge Club garden through a hole in the fence from my old flat, and so I went to have a look. The hole was still there, so I sneaked in. They'd cable-tied a plastic barrier to a piece of metal embedded in the gatepost, and when nobody was looking I cut it with my penknife and squeezed between the barrier and the boundary wall, so that I was effectively about six feet away from the trunk of the threatened tree. The arbs and the SIA goons were not happy about this to say the least, and one of them claimed to have film of me cutting the ties, which I was confident they didn't. I told a goon that it had just snapped. 'How do you explain that?', he said. 'It's like a lot of the stuff you guys do', I said, 'poor quality'. I was soon joined by Jeremy [Peace] who managed to get along the wall and force himself into the tiny gap I'd created, and I can honestly say that I have never been in such close proximity to another man in my entire life. We were so close to each other that we had a chat about taking it in turns to breathe in and out, and at one point I remember creating enough space to raise my arm to take my glasses off and then accidently putting them into Jeremy's top pocket because we were so close.

Jeremy Peace: And to this day I still have those glasses and I still think about that moment we were so close, which is quite disturbing. As Simon wriggled out of the space to go to work, we forced the gap between the barrier and the wall open wide enough for three female campaigners to wedge themselves in. After that, we spent the next hour or so putting on a wonderfully theatrical performance shouting and screaming with pain each time the SIA goons tried to squash us against the wall with the barriers. Unfortunately, we couldn't help laughing in between. They were filming us the whole time, and I think we might have exaggerated the age of one of the female protesters for dramatic effect too. Eventually they all packed up and we saved the Bridge Club for the second time.

Calvin Payne: The third time the felling crews came for the tree outside the Bridge Club, the Heras barriers surrounded the front of the property. Many of the properties on this road had given permission for tree protectors to stand in their garden or driveway to legally prevent felling work. This caused a division between some club members and the committee, when at least one member of the club attempted to enter the property with the intention of persuading others to oppose the felling, and to attempt to carry out peaceful protest action themselves. There was the bizarre sight of the SIA security guards and the police deciding which members could, or couldn't, enter their own club, which we were told was closed for the day. After an hour or so, around twenty members left the club accompanied by police. I don't know whether they were committee members, but it was clear that some sort of meeting had taken place. The tree-felling issue caused divided opinion across the city and that was to be expected, however what made this different is that the SIA operatives and the police were allowing club members with one point of view to enter whilst preventing the legal entry of members who felt differently.

Member of Bridge Club: I said to the police officer, 'Look, I'm a member of the Bridge Club, a keyholder actually, and I need to go in there and do some work for the Children's Club'. The Children's Club is one of our initiatives to attract new members and to create a positive atmosphere between the club and the local community. On the day, I didn't know who the officer was reporting back to, but I found out the next day that the then chairman of the club had already been visited by the police prior to the felling, and had given permission for Amey to set up their barriers on the club's property and also permission to keep the protesters out, including members of the club who were opposed to the fellings, of which there were a fair few. As members we had not been consulted about this. Previously, the then chairman had spoken to me about my involvement in the campaign: 'I don't like what you're doing', he'd said. 'I don't like what you're doing either', I replied. They felled the tree that day, and I remember how sad and angry I felt. The next day, the house manager of the Bridge Club came round to see me at my home. He apologised and gave me a bunch of flowers. He said it was a mistake that I had been denied access. I knew that was a lie, it was a deliberate act and the then chairman of the club had colluded with Amey and South Yorkshire Police. I saw my role in the Bridge Club as promoting the club and the game of bridge in a positive light, and his actions had utterly destroyed everything that I had been trying to do. I never went back.

***

Calvin Payne: This was the biggest attack on a road in Nether Edge and it went on for weeks at a time. There was a fortnight solid when there were crews there every day. It was the depth of winter and during a prolonged cold spell. This was in a neighbourhood where the arrival of felling crews would see scores of local residents and protesters arrive within minutes. The numbers on the street on both sides were as large as at any other time in the campaign.

Jenny Saul: We knew that this was coming, so when I heard that barriers had arrived on Thornsett Road I made my way over there. This was 26th February 2018, at the top end of the road, and as I was one of the earlier arrivals I positioned myself against the wall under a threatened tree. We knew that the injunction the Council had obtained against protesters only required tree protectors to leave a safety zone that has been fully erected. So as long as I leant against a wall between barriers that had not been joined together, I wasn't violating the injunction. I was supposedly violating a minor law, Section 303 of the Highways Act, by obstructing highway works. However, this is such a minor crime that the only punishment is a fine of five pounds per day. More people joined me. Although I had no cash on me, we worked out that between us we could cover it. An 'evidence gatherer' informed me that I was violating the injunction. I told her I wasn't. A security guard told me the same. I told him I wasn't. Eventually, a police officer also told me that I was violating Section 303. I agreed, and offered to pay the five-pound fine. We all had a good laugh as he told me he wasn't allowed to take cash on the street! Then things turned a bit more serious. Which was to be expected given the dozens of burly men in hi-vis, all the police and the tall Heras fencing filling this normally quiet residential street. We were asked to leave twice, and we politely declined. We were told that if they asked us a third time and we declined, the police would give them permission to use 'reasonable force' to remove us. Since reasonable force *should* be proportionate to the crime committed, one would expect very little force for a crime with a maximum penalty of five pounds per day. That's what one would expect, anyway. A masked stranger behind the wall offered to hold on to me to make it harder for the police to drag me off the Heras barrier. Usually one doesn't accept offers of this sort from masked strangers. But this being Sheffield, I knew this guy was a bunny – one of the heroes of the campaign. I hugged that bunny, and he hugged me. And though the security pulled and pulled – four of them – they eventually gave up. The action continued. Despite what was by now a large collection of geckoes, the arborists began hand-sawing branches that did not overhang us. This was a slow process, and we reflected on how long it would take to fell all the mature trees by this method with dozens of security and police in attendance. We reflected on what a stunningly poor use of resources this was for a cash-strapped Council.

Willow: I'd started the day planning for Meersbrook Park Road; a few of us were there waiting for the expected visit from the felling crews who didn't arrive. I used to live on Thornsett Road and so when we heard the crews were there, I walked over with Rob [Pearson]. We both felt an atmosphere of growing tension as we walked up the road, and we started to walk faster and faster. As we passed one of the arbs, he put his fingers up at me and spat on the ground after he looked at me. Rob turned to him and gave him a 'V' sign. The two of them carried on exchanging insults, but I just wanted to get on with the action and carried on up the road. I was geckoing at first, with both me and a SIA bloke in the same very small gap in the fence pressed up against each other. His breath stank and we were very close to each other. I didn't say anything to him. He said a few things to me, though. The tree started coming down and I didn't think we could do anything to stop them. I looked around for Rob and couldn't see him, as it turned out he'd already been arrested for making the V sign! I felt powerless, frustrated at the sound of the chainsaws. I looked at the tree in the garden next to us. I thought if I climbed it, I could jump across to the street tree even though it was around fifty feet up.

Jenny: And then the bunny became a squirrel. He climbed up a tree on private property (whose residents supported our tree-saving efforts), but one which hung out over the fenced-off work zone. He was, however, well above it, so I wasn't sure he had entered the zone.

Yvonne Wragg: I'd given a garden permission for the tree outside my flat, which was the centre of all this craziness with the person up the tree.

Willow: I was wearing wellies and a mask; I could hardly see anything! It was freezing cold and my wellies gave me no grip. I relied on my fingers to grip the bark, which was difficult in the cold. I got to the branch and I started moving along it. There was a point where the branch was rotten, so much so that I could fit my head into the hole! I eventually found myself above the arbs who were at work on the street tree. They were respectful of my tree-climbing ability and told me how impressed they were, while also pleading with me to be careful.

Jules Alexandrou: Everyone was on the very edge of reason. They'd up the ante and then we'd up the ante. It was like nothing that had gone before.

Willow: The police called the Fire Service in order, I'm sure, to get me down and then arrest me. I think they got a 'no' in response at first, but after another call, a firefighter attended. We had a short conversation: 'Are you alright?', 'Yes', 'Can you get down?', 'Yep', 'OK'. It was a waste of their time by the police.

Jenny: I was alternately deeply moved, impressed and scared, for this brave, kind, stranger I had just been hanging onto. As all of us began to appreciate what was happening, the felling stopped. The barriers came down. He had saved the day.

Willow: The arbs had stopped work by then. Amey managers were on site and they didn't want me to slip down the tree, because obviously I could have died, and quite possibly landed on someone too, as there were lots of people underneath me. The police told me that when I came down I would be arrested. Here they were physically endangering my life and they did it on purpose, did it deliberately. I was cold, weak and hungry by then. So

I replied, 'As soon as you go away, I'll climb down'. By refusing they were endangering the lives of me and others. The arbs had packed away their gear and the work had stopped. There was nothing to gain in arresting me by then. The police action was criminally dangerous. I spotted a chance, a gap, and leapt down. I jumped about twenty feet. I'd been up there for three or four hours in the cold and snow. I didn't land very well. There were thirty-two police there. I counted them from up the tree. Some chased me, others spread out to the streets on each side. I was being hunted. I jumped fences, but I'd hurt my ankle jumping down from the tree and I could still hardly see! When I was finally caught I was a couple of streets away and two officers, one a big bloke, about eighteen stone, took me down hard. They just 'clotheslined' me and I hit the ground hard. I was exhausted, that was the end of that. They were annoyed because I'd shown them up, hadn't I? They called an ambulance. I think they were scared of the consequences of what they had done to me. I could have complained but it felt too petty, not the right thing to do. Also I still wanted to keep a low profile!

Calvin: After several hours perched up the tree in freezing temperatures the 'squirrel' wanted to come down. It was towards the end of the day's work. Attempts were made to agree with the police that he would not be arrested if he came down. It didn't appear that anything had been agreed, as when he reached the ground he was pursued by several police officers through gardens and over fences. He was eventually apprehended in a nearby road by at least a dozen officers and had sustained a leg injury in the meantime. It looked like an extraordinary use of police time and numbers for the very minor alleged offence.

Jules: When they were chasing him it was unbelievable! He'd been up a tree and they were threatening him with the Highways Act. When they took him away, he could hardly walk.

Willow: At Shepcote Lane detention suite I refused to give the police my name. They started hassling me to do so. I finally gave my name under advice from the lawyer (which I now don't think was correct). Whilst I was locked up the police went to my home whilst my children were there, which

I wasn't happy about either. When questioned I told the whole story and one of the police in charge said, 'Thank you for one of the most entertaining interviews I've done in a long time!'. I think they genuinely liked me. I could have pressed charges over that day. They might have liked me, but I didn't like them. That day was my crowning campaign moment.

Yvonne: The next day, I finished work early and came home. I couldn't believe my eyes! There were hi-vis everywhere and it looked like somebody had caged my flat. They were felling the tree. I couldn't even get into my home. People were being arrested and then de-arrested. I remember feeling furious, that's why I cried. I told them that they were making a bloody mess! The only thing I was glad about was that they took the tree down to the ground and didn't leave half of it standing, which would have upset my daughter. When I picked her up from school I warned her what we might find outside our house, as I'd seen photos from elsewhere with mutilated trees left in place.

Willow: The second time I was arrested on Thornsett Road it was by the same police officers, and I don't think they realised it was me from up the tree! By then I was prepared to get arrested every day. I'd lost all respect for the police and what they were telling me.

Yvonne: By now I'd decided that I'd had enough – they weren't felling any more trees if I could help it. Someone was there standing under a tree, and someone else asked me if I could park my car under another tree and I said Yes. I was off work because college had closed because of the snow, and my daughter was away. I got quite involved that day but I was mindful that as a teacher I couldn't get arrested. I parked my car under the threatened tree, but I said to the police, 'That's my car, I will only move it if I have to', while I went and stood on a wall under a different tree. Somebody parked their car across the road, blocking it off, and Amey vans were driving at us, which we complained to the police about.

Calvin: Thornsett Road was also one of many across the city that saw work being carried out at the dead of night in an increasingly desperate attempt to fell healthy trees without facing opposition from residents. As a result of this work, the routine of campaigners patrolling threatened areas was extended to begin in the early hours of the morning. The 'cat and mouse' that had previously begun each morning now took place in the dark. When the arbs found a condemned tree unprotected they would use handsaws to hack off as many branches as possible to make the felling easier later on in the daylight.

John Baxter: One night I went out in the area, looking for Amey crews. I was wearing a hi-vis jacket, and realised after a while that I was being followed by a group of tree protesters who thought that I was from Amey!

Jules: Everything was at its peak. They were becoming more and more heavy-handed. I decided to start wearing a mask, but the first morning I went out in it Jeremy Willis [Amey Operations Manager] said, 'Hi Jules!'.

Jenny: It was freezing cold and icy and so when the residents saw that their streets had been gritted, it was a bad sign. The gritting meant that Sheffield City Council and its contractors Amey and Acorn would be back again trying to destroy the trees (which residents had voted overwhelmingly to keep). Thornsett Road is never normally gritted, and all over Sheffield traffic was being entirely stopped by non-gritted roads. They'd also assigned thirty police officers and twenty SIA security that week to deal with us. We were obviously a scary bunch. Can you imagine how you'd feel if you met two pensioners and a fifty-year-old, 5'4" philosophy lecturer in a dark alley?!

Yvonne. I knew as soon as I heard the gritters outside that Amey were coming for the trees on my street. They never normally grit this road, it's not a bus route or anything – I usually wish they would!

Gordon Hawksley: One of my favourite recollections in this whole tree saga was when I got into conversation with a policeman on a bitterly cold day on Thornsett Road. He said that we protesters were very tough to put up with such atrocious weather. He added that he was glad his stab vest was keeping him warm. I asked why he needed to wear a stab vest as we were all peaceful protesters. He didn't answer that specifically but told me they even wore them in the station. I asked why and he replied, 'So that our bosses don't stab us in the back!'.

Calvin: It was coming towards the end of a long, cold winter standing outside in all weathers on the streets. There was one farcical day when we woke to snow that was even heavier, and it was blowing around in a strong wind. It seemed there was no way any tree-felling could be carried out that day, but then there was a call-out: 'Crews on Thornsett Road'! I don't live far away and got there quickly. The barrier vans had arrived at the bottom of the road and I saw one moving very gingerly on the icy road, and out of habit I went and stood in front of it. The police were already present and told me I was endangering myself and others given the conditions. I argued that no-one should be here in the first place and that it was obvious no work was going to be started. Soon there were about twenty protesters present, an amazing demonstration of what we were capable of in the worst of weather, at the end of a very hard couple of weeks.

Chris Rust: A police officer told me that morning, 'Blimey, you folks are tough!'. We certainly were.

Calvin: There was an altercation higher up the road between one of the staff and a protester. Several protesters said that an Amey worker had kicked out at one of the residents who was objecting to what was happening. Despite the number of witnesses, no action was taken in complete contrast to Rob [Pearson] being arrested on the say of one Amey employee days earlier.

Rob Pearson [on a YouTube film]: I was arrested the other day, spent eleven hours locked up, when there was one witness twenty five metres away. Now we've got twenty witnesses standing one metre away and nothing is done.

Calvin: After a couple of hours a police officer announced what was obvious from the beginning, that no work would be carried out that day due to the weather. My day was improved by seeing Darren Butt of Amey stranded just around the corner in a car that wouldn't start! 'Morning, Mr Butt' I said as

I walked past on the other side. 'Thank you, Calvin', he replied, somewhat tongue in cheek.

Yvonne: Throughout all this I met a lot of new people, and there was a real sense of community. People that I'd never met were hugging me that day. I really love trees, whenever I need cheering up I go to the woods, but this became about so much more. I won't vote Labour ever again, it showed me how corrupt they were. I lost a lot of trust in the authorities that I thought were there to protect us. I don't trust the police, and for me it came to be about more than just the trees. It was about money, lies and corporate greed.

2017-2018

'I remember overhearing them say, 'This will be one of the roads that decides it' and they were certainly right about that!'

Maureen, Resident

Meersbrook Park Road is a tree-lined street running alongside the park of the same name. It is adjacent to a busy air-pollution blackspot.

Maureen: I live on Meersbrook Park Road and when the Council brought in the crew to assess the tree outside my house I was watching through the bedroom window and they were writing something down about my tree. I came downstairs and I went outside and I asked them, 'You're not taking this tree are you?', and they said, 'Well, not yet, but it will have to come down'. I asked them why and they told me, 'I'm afraid it's a danger to your property', and I just laughed. I said, 'Look, I've been here sixty-three years and my house is still standing and there's nothing wrong with it. This tree is perfect and I've seen it grow from a very small, skinny tree into this fantastic specimen. You're not taking it!'. Then this young woman came over and said, 'Well I'm sorry', and I said, 'You're not having this tree. Over my dead body you'll have this tree'.

Calvin Payne: This was a typical first experience for a lot of residents early on. This was before any signs were up or any plans were made public. This was when they just turned up with the list that said, 'Your tree is coming down'.

Maureen: I got in touch with Paul [Brooke] and we all got together and agreed we wanted to keep our trees. And then it sort of flared up! We felt very angry and there was disagreement about what to do, letter writing and petitioning and what have you, but what really happened here, as much as it did many other roads, is that we were doing *something* about it, just as we always knew we would.

Ruth: Back then, we always said 'our tree' on 'our road', we didn't realise it was the whole of Sheffield involved. We decided, 'Well, if we have to, we'll tie ourselves to it', but then we started to realise it was bigger than that.

Maureen: I was shocked that they could just come onto our road and absolutely slaughter the lot.

Ruth: All eighteen of them!

Maureen: It was a big steal, they thought they could just come and take them all away and I couldn't say no. My husband said to me, 'It's no good you know, you can't fight the Council', and that's when I said, 'We bloody well will fight the Council!'. 'You won't win', he said, 'you're wasting your time'. 'You just watch us!', I said. I had no idea at the time about what could happen, about police involvement or anything else.

Ruth: So Mum tells me if this is what happens, then this is what she's going to do. We agreed that we'd guard the street, and then we realised it wasn't just our tree, it was eighteen trees. And when they first came, that's when we all just started standing under each other's trees.

Maureen: And we put alarms on to wake up in the morning, so that they couldn't come at two o'clock. Like they did on Rustlings Road. The first day they came was actually very gentle compared to what happened later on.

Ruth: Was that the park? Was that when they tried to put barriers up in the park?

Maureen: It was the day they filmed us, trying to tell us we were breaking the injunction in the park.

Calvin: My and Alison's court date was already set by then and the Council people turned up and read a version of the injunction which left out the crucial word 'highway'. It was bizarre, as we were all in the park and knew the injunction didn't apply there. A few of us were shouting back at the process server and for some reason I had a mask of my own face on the top of my head which probably didn't help my argument!

Ruth: Because we all just went into the barrier inside the park confident that we weren't doing anything wrong with the law.

Simon Crump: There was a very surreal moment when a bloke with a sports bag turned up, put on a protective mask and produced two rapiers. It turned out that he was a fencing instructor on his way back from giving a private lesson, and he offered us one for free. There's a great photo of Dimitris [Diamandis] having his lesson inside the barriers in the park, and when he took his helmet off at the end, with his little moustache and beard he looked like one of the Three Musketeers!

Ruth: Paul pretty much explained the injunction to us that day. The park was not a highway so we knew we were alright.

***

Simon: They concentrated on other streets after the fiasco in the park, but then a few weeks later they came back with a vengeance. At the end of the first week of real trouble on the road, about twenty of us met up in the communal hut on the nearby allotments. The first thing I recall was

Rob [Unwin] walking in with a bicycle lock in his hand, putting it around his neck and then locking himself to an upright part of the shed. 'Well, that seems to work', he said. Various people seemed to be carrying, and experimenting with, various items of borderline bondage gear and I'd just bought a pair of handcuffs off eBay, proper ones, not the fluffy pink sexy ones or anything. So Susan [Ashworth] and I spent the entire meeting handcuffed to each other. Which was nice.

Susan Ashworth: Except I couldn't reach the biscuits!

Simon: Paul was chairing the meeting and from the way the other residents were looking at him, and the way they were talking to him, it was very clear that he was in charge and that if he'd have asked them (very politely of course) to do pretty much anything, they'd all have done it, just to please Paul.

Susan: Yes, Paul was fantastic that day!

Simon: What was really upsetting here, however, was the nagging thought that if Paul, or someone like Paul, had lived on Rustlings Road then the whole terrible business there might never have happened. Halfway through the meeting, there was a knock on the door and an allotment holder came in to make a cup of tea. Lord knows what he must have thought, that maybe we were some weird geriatric sex club or something. Anyway, Paul had made these devices out of sections of metal pipe, which you could put your arm inside and pretend to be locked on to the park railings. The whole meeting was like some strange amalgam of *Last of the Summer Wine* and *Dad's Army*, but above all it was very moving. These people, these residents, loved Paul nearly as much as they loved their trees. And there was absolutely no doubt in their minds that they were going to win.

Rob Pearson: I'd just moved back to Sheffield and I was walking down MPR when I saw Ruth and Maureen getting man-handled. Ruth was shouting because they were twisting her arm back, and I was asking, 'Hang on, what's going on?', and they told me what it was about. I love trees anyway so it wasn't just a case of just not letting these women get hurt. First thing I said to Ruth was that I promised I'd come out and stand with them.

Ruth: And he did. Every day!

Rob: It was the least I could do. They were bullish at the time when I walked by, being pretty bullish. All the hi-vis too, which I didn't like. And I thought, 'Well I ain't the hard man, but I'm not having this'. And the love of trees pulled me into the campaign as well.

Helen Damnation: My first action on MPR was when I watched Alison [Teal] standing in a front garden under a tree that was barriered off, and a young boy who lived there sitting on his wall. The crew carried on working, spraying them with sawdust. I was incensed and vowed to take whatever action I could. The next day, I was ready with a disguise, a scarf across my face, as I walked down MPR as the barrier vans were just arriving. It was a cold day, and so I think they assumed that I was a passer-by. As they started erecting barriers, I stood under the tree they were sectioning off and stayed put. That was the day 'Santa' came to the street; he stood under the tree we called 'Stumpy' [a tree by the park gates which had been previously attacked], and despite being surrounded by security guys all shouting at me to leave, I felt safe because of all the supporters around the edge of the fence, passing cardboard and magazines, but more importantly their support for my actions. I subsequently supported many other bunnies, helping them over the park fence, passing them hot drinks, and when not in disguise I sang and bantered with the SIAs and barrier guys, and helped block the vans.

David: Maureen's husband [Gordon] ended up getting involved after he said he wouldn't at the start. That was brilliant. He was great, really great, standing out there all day sometimes.

Ruth: We kept walking out and leaving him in the kitchen on his own. Six hours later we'd come back all covered in snow. And then one day he actually said, 'Right, I'm with you'. And he was!

Rob: He got there in the end didn't he? Top man!

Ruth: He said, 'OK, I'm with you, I'll get ready'. And he got ready that morning and I said, 'You better wrap up', because he'd been led to believe because Mum and Dad are in their eighties they wouldn't touch them. It wasn't until the week when it all kicked off and the fences came when he was moved by the SIAs. They said, 'We will remove you, but we'll get a woman to do it'.

Maureen: I did say to Gordon, 'You know, it's good for you to come out, because if they do remove you it will look really bad, because you're nearly ninety-years old. They said, 'We may be a bit rough', and I said, 'Don't worry about it, I don't mind'. Gordon said, 'I can take it'. So I think at first he thought he wasn't going to be useful enough, that he couldn't do much. I said, 'They might hurt you or push you around', and he said, 'Let them do it. I don't mind'.

David: Just being on the street, just showing up is sometimes enough. Showing your face. Any support is better, because behind closed doors it's not going to do anything. And remembering my little bit, you know, being attached to the fence, there was my neighbour, who lives opposite me, who had only lived in that house for three years. We'd just said pleasantries to each other, we didn't know each other that well. He was having a day off and he came up to have a look at what was happening on Meersbrook Park Road, with his teenage sons, and just ended up getting embroiled in it. He was actually on the other side of the fence with his kids in the park, grabbing hold of me!

Fran Grace: I'm a retired primary-school teacher, hardly a troublemaker at the best of times. I promised my daughter and my little granddaughter, who lived on Meersbrook Park Road, where so many trees were due to be felled, that I would do my best to protect 'their' trees. I didn't want my granddaughter to ask me one day, 'What did you do to prevent this mass destruction? Did you care? Or did you stay at home, thinking that others would go out instead?'.

Willow: I used to go for a run around the park every morning. I'd been aware of the campaign and I'd been self-employed, I was always working looking after gigs. I didn't really have any time to do anything, which was just a growing frustration for me. The felling crews kept coming in, trying a little bit here and there. I think they'd come in and trim some of the trees [on Meersbrook Park Road], they'd come in the night and chop some branches down and leave them in the park. It must have been October or November 2017. I'd been for a run and I had my running kit on. It wasn't warm-weather gear. They were trying to take down 'Stumpy'. As I walked past, I noticed a couple of people standing by the tree, including Phil from the bottom of the street, and I stood with them. Then they put the barriers up around us! So they made the first move. I didn't make the first move, I was just standing on my street and they put the barriers around me. Maybe it's because I don't like bullies and I don't like being pushed around. That was my stand after that, 'You put the barrier around me, OK, fine, well I'm gonna give you two back'. They started to come in and I just climbed up the tree. I just whipped up it. I didn't know about the injunction. I remember distinctly that I did have a really big cup of tea before I left the house, and that kind of haunted me for the next five hours in that tree; I've never needed a wee so badly! The people on the street were amazing. Everyone was calling up, shouting to me, 'Are you alright?'. Even people I didn't even know – although most people did though. Most people had the sense not to call me by my name!

Jimi: That was the beautiful thing about when you were in that fence and you didn't know what was happening, you'd sit there and you would think, 'Oh my god, I'm here'. My main concern is to look after this tree I'm leaning against, I don't really know what's going to happen. And someone would shout, 'You can't touch him!' and so you think, 'Oh they can't touch me'. Someone would shout, 'Here you go, mate'. You'd look and someone has thrown you a pillow or a piece of cardboard or something to sit on. Then you'd realise that you weren't alone and people were there to help you. That meant a hell of a lot when you were there on your own. We've all been there at some point, feeling like we're on our own.

Rob: I didn't know anybody before. That's the beauty of what's come out of it, a lot of friendships. Another fond memory that springs to mind is the handcuffs. I don't know which lady it was, but I was there early one morning and she just turned up, and it's a big box of handcuffs. She just started handing them out and everyone is giggling and thinking, 'Yeah, we'll use these in the bedroom', and nobody's handed them back. I've still got mine.

Ruth: I've still got mine!

Willow: Another bunny called me by name inside the barrier with all the cameras on us and I covered it up so badly by drawing attention to it. I wasn't masked or anything. I even took a selfie. I got a message from my girlfriend at work saying, 'What are you doing, Willow? You do realise you're not going to get away with this'. But I was just like, 'Fuck it, man'.

Rob: Kinda my thing was to stand in front of vehicles; I quite enjoyed that. I'd never done it before. It just seemed really easy. I enjoyed it, within reason. I've got a right to do that. Everyone had their specific job, like Jimi, he had the trees covered and I had the trucks. It's all about small victories. They used to really piss you off and get under your skin, and the one I remember for me personally is taking a tenner off a hi-vis guy that had been pissing me off for a time and I had a bet with him. He told me, he came out and said, 'You won't get more than ten minutes in front of the van tonight'. I said, 'I bet you I get twenty, mate', and he goes, 'I bet you don't'. And I said, 'I'll bet you a tenner'. He said 'Alright' and I shook his hand. I got my twenty minutes. He lost and the next day I shamed him in front of everybody. He came from around the corner, saw me, gave me the tenner and I bought a 'Tree Hugga' tee-shirt with it! I told him that's what I was gonna do, the proceeds are going to the campaign.

Jimi: Well, I was up a tree one day and I was thinking to myself, 'Is that vehicle gonna go so I can climb down so they can't see where I get my ladders from?'! It was well past 4 o'clock.

Maureen: We hid the ladders in the park!

Jimi: I started off dressing in different disguises and got heavily beaten up in Dore, on Chatsworth Road, when they had come down and put a barrier on my shins when I was trying to go under. So, when I came back to Meersbrook Park Road the next day, I was talking to a security guard, holding my nose to disguise my voice, and he didn't even realise that he was talking to the same person [as] the day before. And I was saying, 'That was really nasty what you done in Dore', and he said, 'Oh well, he nearly took my head off!'. Anyway the next day I'm on Meersbrook Park Road, talking to that same guy whose hat I'd knocked off the day before, when he said, 'He was a really big fella, too!'. Well, I'm not a really big fella, but I had to crouch down a little bit and say, 'It wasn't me' and because I was wearing a different disguise, he believed me and I got away with it. So then I decided that to get in closer, I'd wear hi-vis, too, and a cowboy hat, and so it worked. Because I could get in a little bit closer wearing hi-vis and the cowboy hat was just my joke about them being like cowboys, and it became a bit of a trademark for me as well.

Ruth: So it was Jimi who started the whole 'Shamey' thing off?

Jimi: I'd already seen the word used on Facebook, I think. So, yeah, I spent a couple of days jumping in and out of the barriers dressed up as a cowboy. On one occasion on MPR, I saw that the bunny that was in there already had clearly damaged three barrier fences. Everything had stopped and that's what drew my attention to it, and I thought, 'Well, what's going on?'. And I realised they were waiting for the police to arrive and arrest this bunny for criminal damage to the fences. So, I go down, and I've got a knack of being able to lift these fences up and part them and be able to get through them without damaging them, which I did, and I was in. So I got chatting to the bunny after and then said to them, 'You know, if you don't get out of here, they are going to arrest you for criminal damage on these fences'.

Calvin: Those Heras barriers bent surprisingly easily at the top. Occasionally they got helped to do so by the SIAs and the arbs. Some of them pointed out, 'I think it's bent there', and tried to do one of us for it.

David: Some of them have bits of steel sticking out anyway and they weren't in good condition.

Jimi: So he managed to get away, then they arrested me, then the evidence gatherers showed the policeman the footage. Clearly the person wasn't in hi-vis who bent the things, so they de-arrested me. That was a first. I've never been de-arrested before. They didn't put me in the van, but they took me to it. I put my arms up and let them search my pockets.

Calvin: After that there were loads of de-arrests. They were doing it almost every day. I'd never heard of it before. People were taken away, read their rights and as soon as they'd take their name and address they just let them go again.

Jimi: I realised very quickly that the best thing to do really, was to just get up a tree and hold that tree. So then I started turning up early in the morning with a ladder and local residents here, people on Meersbrook Park Road, helped me get up the tree, and then took the ladder away and hid it.

Simon: I think that MPR in particular was the closest we came to operating like the anti-frackers. I went up to Preston New Road [fracking site in Lancashire] one day, and after they'd all stopped taking the piss out of the way I was dressed, we agreed that the logistics at least of their campaign, were more straightforward and easier to plan for than ours, because they knew exactly where the fracking was going to be happening every day. We never had that, we never knew where we were needed, it could have been anywhere on any day, and several places on any given day, but in the final weeks of MPR, when they really got stuck in, it was a pretty sure bet that they would be there, day after day. So in a way, I think that was a failing on their part because it made it easier for us to mount an effective defence and put people up the trees. Those ladders were bloody heavy, though!

David: Because of our previous experiences with residents in Norton Lees I was wary about passers-by on Meersbrook Park Road. I was worried I would get abuse. It was such a change when people went past and were being supportive.

Maureen: On Albert Road they wanted to pull your mask off, didn't they, David? And we said no. We stood between them – they're not going to touch our David. Over our dead bodies they will.

David: I watched back footage on YouTube recently, of the day when Ruth and I were both crushed against neighbouring trees. That particular day, we thought we'd got it covered. We were both inside the same barriers, because it was a single circle of barriers wasn't it?

Ruth: It was a really long work zone, it covered quite a few trees.

David: I guess that was the day; well maybe there were some previous times too, where I knew that I'd be willing to go that extra mile. And then it was like, 'Alright, OK'. It looks like today is going to be an interesting day. Then I clamped myself to the barrier. They attempted to forcibly remove me. I'm happy to say they didn't succeed. We were threatened with court action. There's a piece of YouTube film where Paul calls out Darren Butt for wrongly quoting the injunction to us.

Ruth: There had been four of us under a tree and they prised three of us off the park railings. I was the last one. He tried to get me off, but I hung on, and I said 'I want you to prove we are breaking the injunction'. And he said, 'I've definitely got enough evidence to take you to court for breaking the injunction. Do you want to risk that?'. And I said, 'Well, I want you to prove to me that being on the wall, holding onto the railings, actually means that I'm on the highway'. I told him, 'I will gladly get down if you will prove to me that I'm breaking the law'. They went away and then he came back and said, 'We haven't got that information', and then tried to prise me off, unsuccessfully. But then eventually, unfortunately, and this is one of my biggest regrets, I had to get down because I didn't think I could afford to pay the fine, but that was after Paul [Brooke] was saying they'd pretty much lied to us to get us out.

Calvin: Well 'we haven't got that information' is a funny way, a good way of saying it doesn't exist.

Ruth: It made me laugh because one of the things he said was, 'Do you know that your insurance will go up because of this, because of the police involvement, all your insurances will go up. And look, you've got this park to look at, you don't need these trees'.

Rob: It was always forever unfolding, wasn't it, because at one stage we believed it to be 6,000 trees to be felled and then you actually find out it's 17,500. Fuck that!

Calvin: I remember Jack Scott [Labour Councillor and former Sheffield cabinet member] saying in an interview, in 2013 I think, that 1,250 trees were to be felled. From there the number just seemed to go up and up.

Rob: I mean that was the point where I said, 'I'll stay here and see this to the end'. How can they possibly do this and get away with it?

Helen: When I was inside the barriers, well, I just created a wall, you know, for when they're all gathered round you and shouting in your face. When Simon was arrested for sitting down in front of the vehicle on Meersbrook Park Road, me and Jimi were at the back, but we were in disguise because we'd had loads of threats that day. You could see how wound up that copper was!

Calvin: I was further down because we were blocking the road at each end. Russell [Johnson] was crossing the road doing the slow walk, and Russell is as polite as they come. The two cops that got out of the car, the woman copper (it was a young guy that lost his rag with Simon), she jumped out of there, marched up to us and just said, 'We're going to arrest you', to Russell, when in fact he was already getting out of the way, so I thought, 'Blimey, she's already decided here'. So I rushed up to Simon and said, 'You'd better know what you're doing here, because they're on one today. They're going to arrest you'. They definitely came with a preconceived idea, didn't they?

They didn't apprise themselves of anything first.

Jimi: I'm right beside them. I'm standing there in between them pretty much and the copper's getting more and more wound up, you could see it, and Simon's not helping the situation.

Simon: As always in these trying situations, I was being the master of diplomacy!

Jimi: Anyway I thought he might get a bit heavy-handed, especially when he grabbed Simon by the arm and he tried to yank him up, because that's not right at all. Especially with the rigid cuffs that dig into your wrists.

Simon: They fucking do! They definitely weren't the fluffy pink sexy ones I was hoping for.

Jimi: I did actually say, 'Go easy' to him. I don't normally say anything obviously when I'm masked up. What I did was say, 'Go easy', and then put my hand out in a Star Wars manner.

Simon: Paul had told me (and obviously I was always going to go with whatever Paul said, because I viewed him as my responsible adult at the time – and still do actually) to wait until the third and final warning, we were expecting the 'Lionel Richie' approach (you're once, twice, three times going to be arrested), before I shifted out of the way and then someone else (out of earshot of the copper's warnings, obviously) would take my place and so on. The plan was that as a team effort we could delay the felling crew all day so that they couldn't work on MPR, and crucially, not allow them to go anywhere else to fell trees either. Only it didn't quite fucking work out like, that did it? First thing I knew about it was Calvin telling me it was a young copper and a female copper we hadn't seen before, and that they were already not in the best of tempers. 'You'd better be ready to be arrested', he said, and asked me, 'You seriously up for this?'. It was cold and my bum was wet from sitting in the road. 'Fuck it', I thought. 'Why not, eh?'. They didn't fuck about either, not the usual three warnings or anything, they just grabbed hold of me, handcuffed me and marched me to the cop car. If you watch Nadeem [Walayat]'s video, the copper asks me why I'm not going to move, and I say, 'Because you're annoying me and I can see right up your nose'. Which in retrospect I probably shouldn't have said. It didn't exactly contribute to the carefree mood of the day, to be fair. As they led me to the car I said to the copper, 'You lost your temper just then, didn't you? That wasn't very professional was it?', and he lost his temper again. In the car, on the way to Shepcote Lane custody suite, he was complaining about all the paperwork he was going to have to do as a result of nicking me and how he wouldn't be able to play football with his son now. 'Drop me off here at World of Sheds, then', I said. 'It's my favourite place in Sheffield'. I probably shouldn't have said that either.

***

Rob: It was half term when me and Willow walked over to Thornsett Road, the day when we both got arrested. Fifty cops there, and I was like, 'What the hell is going on here?'. That was a change and I think Meersbrook Park Road that day was the change, when they really upped their game.

Helen: It's when they started carrying bunnies out of the work zone. Bunnies had sat there until then and they left them alone, and then all of a sudden they started carrying them out.

Rob: It was like a moral victory for us on MPR and then they thought, 'No, we're upping the game'. Fifty cops to a tree, it doesn't seem possible looking back on it! The size of what we were up against, police, SIAs, everything. I remember bending the thumb of one of the hi-vis guys as they were doing the same to the women on the park railings. Well, I got hold of his thumb and I just leant my weight on it and he just yelped like a dog. 'It hurts, doesn't it?', I said. You know what I mean, I just let him know how much it hurts.

Jimi: Then they came unstuck because they thought that the threat of arrest was gonna stop us. So, I purposely went and got arrested. When I was being carried, one of the coppers who had my arm was doing that grip with his hand, really nipping me under the arm here, it really hurt even through all the layers I was wearing. Because I'm strong enough, I just brushed his arm off and gave him a look. And he came back, next time he wasn't giving me nothing. I wasn't struggling, I wasn't kicking out. I was wearing steel-toe-capped boots and I could have made it really difficult for him, if I wanted to. It wasn't about that though, was it? Of course they wanted me to. At Shepcote Lane, they sat me down and they said, 'Give us your name and we'll let you go'. I didn't say anything so they said, 'You have the right to remain silent', and I thought, 'Right, *silent*, I'll definitely stick with that one!'. So they had to release me in the same cowboy gear that I was arrested in and I walked all the way down the canal back into the city dressed like that.

Rob: After my arrest, I had to walk home from Shepcote Lane and it was midnight, so I sent a message to STAG via Paul [Brooke] saying, 'Everything's cool and thanks for the great support, but it would be nice if someone meets the arrestees when they're released'. After that, it changed straight away. If you got arrested they'd always be there to pick you up. That was important, to have someone waiting and not to have to find your way home on your own. I'm thinking, 'Am I all alone in this or don't people care?'. But they did care. As soon as I told them, they rectified that, the campaigners, and they always had someone to come and meet the arrestees when they were released.

Jimi: When I was arrested, I didn't have my phone on me. I didn't want to take anything with me. The only thing I took in my pocket was a crystal. A quartz crystal. The desk sergeant didn't even know what it was!

Helen: When I was arrested with a crystal in my pocket, because I wouldn't give my name at the police station, that's what they called me, 'Crystal'. Well I've never been arrested before in my life. So that was a first and it wasn't such a bad experience. They look after you, they feed you. You've got a bed and toilet.

Calvin: I told them I've been in worse backpacker places as well. With the arrests, people that haven't had it happen to them used to think it either went one way or the other. They either thought it was dead cushy (which it wasn't) or they thought it was worse than it was. It was frustrating and boring. The worst thing is when they just threaten to arrest you. Well in that case, you make them do it or make them back off. You aren't going to lose to just a threat. I was determined to make them actually do it. If they beat me by doing it, that's one thing. It's like the injunction in that regard. Don't just come out and say, 'You can't do this' – you have to enforce it.

***

Rob: Another good memory for me on Meersbrook Park Road is in the dank, wet, sub-zero snowy days, where Benoit turned up and sang us a song. If it was out of tune, it didn't matter.

Calvin: He sat on the floor in all weathers, didn't he?

Benoit Compin: I'd been seeing it happening all over Sheffield, but I was moving around, living away, in between things. I'd been to France quite a few times, so I wasn't really in Sheffield anymore, and I was wondering what would be my next challenge. I was prepared to go back home or do other things, whatever. I was in France, I think it was during the Rustlings Road felling. I watched this from France and I was, well, there are no words, I just couldn't believe it. I just couldn't. I thought, 'Do I really want to go back to Sheffield?'. But then I did come back, it was really taking up my thoughts when I was in France, and when I came back this was still going on. I lived on Meersbrook Avenue nearby, and I was at home going, 'Oh my god! Oh my god!'. So it was happening, and on the road just down from me. There was a video on the STAG Facebook page, which I had joined, of women shouting and bouncers trying to take them off the fences, and I was like, 'This is not happening'. So I just came down to see, I was really curious, and the first thing I saw was the SIA, 'Bob' I called him. Inside the barriers, he was next to a tree, and I came to him. I asked him, 'What's going on here?'. And he's just trying to act on the defensive, going, 'Oh, I'm just doing my job, there is nothing you can do about it'. I thought, 'Whoa' and I said, 'What do you mean there is nothing I can do about it?'. I started questioning him and, yeah, I was just not having it.

Rob: Benoit was funny and he really got under their skin! He was definitely, definitely, on board from the off.

Benoit: With my magic I couldn't do anything. I'm a close-up performer so I thought, well, the role of a poet is not so different, because that can sow spells in a way too. Words create spells, and they do have a power on people. They can create the magic. I decided to experiment with that. I really tried to embody the character of the poet. I was working at the time, and I was trying to find inspiration in animal spirits, for example, and trying to draw back to, for example, the crow. As an animal it's got quite the mythology. I thought that's the effect I want to create on the workers. I wanted to look like a crow. Nobody likes a crow's noise. There was one worker who couldn't stand being next to me because he couldn't stand those words. And once I had them rehearsed a bit, once I had the poems and the rhymes and I had them in my head going, I could eventually just get started and bring the words out. There were a few I used in different places. The 'Bob' one, 'The face for the job', was popular, but then I got a bit tired of the poetry and I then was trying my voice and I thought, 'I need to make more noise'. Even looking ridiculous because I did not have a guitar at that time!

Calvin: One of the things Benoit was in court for was going in and saying he had been booked to appear: 'I'd been asked by these people to come and perform!'.

Rob: On a dull day when you'd feel defeated, Benoit would turn up round the corner and sing you a song and it's like, 'Yeah, it's alright. Everything is going to be alright'.

Jimi: I climbed up a tree on Meersbrook Park Road when they didn't even turn up for two days. Just had 'em drive past. I was walking down there with a ladder on my shoulder with Paul and the kids were all looking at us and we're in full gear with the cowboy hat on, and he goes, 'Don't worry, kids, it's dress-up Friday!'.

Rob: Paul used to come past on his way to work and wind the window down, just to check on the trees, to check it's all in order, and we're like, 'It's OK, Paul, we've got it'.

Simon: There was something about Paul which inspired us all, and he won't like me saying this, but people show their true character under pressure. Paul's a born leader.

Jimi: We were the Meerkats!

Ruth: Everyone was looking out their windows in the morning, weren't they? We were ready.

Maureen: One of our group, Janet, was our early-warning system. She said, 'Right I can see the lights on the main road now, they're coming'. And we'd ring one another and say, 'They're here, they're definitely here'. She hid the ladders down the side of her hedge. The following day they came and searched the park for ladders!

Jimi: Because they thought we were hiding them in those bushes in the park, didn't they?

Maureen: They've cut those bushes down since. They've cut all the bushes down. That was a great place for people to change outfits in!

David: I remember as well the day the barriers came down, we all stood with the tree and looking back on that image, I remember thinking as well at the time that everyone is masked up! I'm the only one without a mask! And they're filming me!

Ruth: Chris and Jan, that had just moved onto the street, they'd only just moved on that day and were asking, 'What's going on?'. So we told them straight away.

Maureen: They were still living out of suitcases. They came down and they said, 'We're your new neighbours and we're with you'. And that was it. It was so good. And immediately they were family.

Ruth: Remember when they used to drive down the road when we were all standing out and we always got people tooting, didn't we? People were so encouraging.

Rob: We did have it sewn at the top, the middle, the bottom. Everyone knew exactly what they were doing.

Jimi: And those park railings. We were like limpets on there!

Ruth: From seven o'clock and sometimes a lot earlier than that weren't we?

Helen: We used to do the three till seven shift, didn't we, in the mornings?

Ruth: Jimi stopped them twice, didn't he?

Jimi: Yeah, I stopped them twice at three in the morning!

Calvin: Night raiding, yeah. We used to drive them mad when we were there at that time in the morning. The look on their faces. They couldn't believe we were out there. They knew they were in trouble when they couldn't get out and work at that time!

Helen: I think one of my favourite memories is when they'd got the Heras barriers out, outside the dog-free area in Meersbrook Park, and I just stood there against the railings and bunnies were hopping in over the fence, using me as a bunk-up. They were getting dragged out and they were just going round the park and back in again, and back again and again. It was so lovely.

Simon: It was wonderful to see! It was like some sort of bunny-based relay! It was really inspiring, and I think it was later that day, inspired by all that, which finally landed me in the High Court. I was outside the barriers with Fran [Grace] and they put another row of barriers up behind us. Then they told me that I was inside a work zone and that I was breaking the injunction. I think I just told them to fuck off at that point, I was just so sick of it all. I knew that they were going to get me, and I had already breached the injunction, on camera, at least twice before then. Anyway, thanks to Calvin, it had already made it easier for the rest of us to do that because we'd all seen him deliberately break the injunction and not get sent to prison.

***

Calvin: It was of course on MPR where the decisive battle was fought and won. The photographs of a dozen or more people around a tree, all in breach of the injunction, many unmasked, spelled the end of that legal device. It had been defied and challenged for six months by then, but never so decisively. This wasn't one or two people being difficult. It was a protest and a celebration in one, and it is very appropriate that it happened where it did. That's the moment when the injunction became worthless, and without it they couldn't beat us.

David: We're in! Someone shouted 'Everybody in!' and we did.

Ruth: Is that when we started singing 'Happy Birthday' to somebody as well? It was one of the bunnies' birthday, wasn't it?

Calvin: That bit of footage, I think it's a bunny, a very mild-mannered bunny as well. I think I know who was shouting, 'Everybody in'. Then the fences are on the floor, one of them is a bit buckled, and so our people just go in. It was the dam finally breaching. From one or two bunnies to a few and then to all those people at once! The culmination of six months of us versus the injunction.

Ruth: That was such a wonderful moment!

Nicola Gilbert: I was there with my husband, Richard, when it happened. None of us that went in the work zone did it in a sneaky way. We did what we passionately believed in.

Richard Gilbert: It was freezing cold. Benoit was reading and playing the guitar. It was the only time I went inside a work zone. I just thought I'd better support Nicola; if she's doing it I can't really just walk off, can I? There were so many of us that they had to read the injunction threat over a loud hailer! Everytime they read it out the women just made a noise and shouted back. It wouldn't have been difficult to find out who we were. The photo was in the papers. They just seemed to have given up. One of the blokes taking our pictures asked me, 'Why haven't you got a disguise on?'. I said, 'Oh, are you supposed to?'. I had no plans to do that, I was there in a supporting role as it were. When the barriers went up I thought, 'Time we were off', but then all these folk went and gathered around the tree and I said, 'Oh well, I'd better join in'. No time for any consideration!

Nicola: I'm not afraid to say that's me in the photo. If there are repercussions now, then that's absurd. I haven't got small kids or anything to worry about. If I was in jail for a few months nobody would really mind!

Lisa Fletcher: The battle took place right outside my house! It was my son's fifteenth birthday and he was poorly in bed with tonsillitis. He kept waking up throughout the day with a high temperature asking 'Why do they keep shouting "Shame on you!"?' and, 'Why are the police pushing everyone around?'.

Calvin: Ultimately someone in authority, somewhere, must have decided that the game was up, looked at the mass breach of the injunction and said, 'This is it. We can't beat them'.

David: I knew something was going down that day because I remember seeing Paul Brooke's face and kind of nodding to him. Just the look on his face, he was distressed because of what was going on inside the barriers where a female bunny was being roughed up. I myself was kind of, just by chance, right next to him when he started to push. Paul was right there when the thing started to go. Instinctively, I was like, 'Right, I've got your back, I'm moving with you'. And then, because there were about fifteen staff in hi-vis who started to push back against him, I remember putting my shoulder against the fence.

Calvin: He [Paul] backs into me, then backs into the fences and security guards and they all push back against him. I'm there as well because when they all pushed back against him, that woman, the evidence-gatherer supervisor, says to me, 'We've got you now' because the fence hit into me and they said, 'We've got you for criminal damage' because Paul pushes forward, and they push him back into me. When Paul was charged with breaching the injunction I played an inadvertent part. I didn't realise that our conversation on the street was being recorded by an evidence gatherer. I told Paul that they were 'really on one today', and that one of the SIAs had already 'beaten someone up' that morning. Paul was, of course, cleared as he proved that his actions were in order to go to the aid of a bunny, and meanwhile the Judge said that I was 'somewhat prone to exaggeration'!

David: At that point it's just your instincts kicking in. It just felt to me that everything was going to collapse.

Calvin: It felt to me that if they came back the next morning, then anything could have happened. There could have been hundreds of people there, could have been a huge number of arrests. Someone decided to pull the plug. Probably a late night meeting somewhere with the Council and Amey, maybe the police too.

Ruth: We'd had a meeting and we were so ready the following morning!

Dave: The day that never was. What would have happened that day?

Ruth: We had special devices that had been created for defence. All ready in people's gardens. We had the lock-on tubes...

Ruth: We were a bit gutted that they didn't come that day, we were just so ready.

Calvin: I mean anything could have happened that day. That was my feeling on the previous day. The stakes had been raised beyond anything that had gone before. A lot of people had shed their inhibitions over the previous couple of days. And that's when the Council can't win. They can't win at that point. When we went through the whole campaign with a small number of people prepared to do certain things, and then it just sort of grew bit by bit.

Ruth: Yeah, people got braver and braver as it went on.

Calvin: Then suddenly you think everyone who's there is going to be more or less ready to do something. They've come this far.

Helen: The thing that helps it feel like that, though, I mean I didn't jump through a fence or anything like that, but they'd put the barriers around me. They just thought I was walking down the road with my shopping or something. Haha! I was ready, I'd got cushions, I'd got everything ready.

Calvin: I remember when Nether Edge was at the centre of the tree campaign, and I remember the very early days when Rustlings Road was too. I said to Paul, I think in November or December 2017, that Meersbrook Park Road is now the nerve centre of the campaign. It was appropriate that the campaign didn't just fizzle out. I don't mean fizzle out in defeat, I mean fizzle out and we win. Not with one day when they could just announce, 'We're stopping', but it was really appropriate that it ended in an unprecedented action. So that we can say, 'That's why it stopped' – so we can show that. They can't get away with saying, 'Oh, we decided to stop'. They were forced by the people to do what they did. I'm glad it finished with a bang, and on a brave action. And now you've got the simple, cheap and easy road repairs being carried out on MPR and everywhere else. Everything we said could, and should, have happened all along.

Maureen: I remember overhearing them say, 'This will be one of the roads that decides it' and they were certainly right about that!

CHATSWORTH ROAD 17	OLIVE GROVE ROAD 2	DUNKELD ROAD 11
LISMORE ROAD 8	RAVEN ROAD 7	SHELDON ROAD 7
VAINOR ROAD 6	WOODSTOCK ROAD 7	WAYLAND ROAD 11
MARDEN ROAD 7	KENWOOD ROAD 7	RAVEN ROAD 7
CHIPPINGHOUSE ROAD 7	RUSTLINGS ROAD 11	BANNERDALE ROAD 7
DUNKELD ROAD 11	WAYLAND ROAD 11	ST. RONAN'S ROAD 7
THORNSETT ROAD 7	MEERSBROOK PARK ROAD 8	SPRING HILL ROAD 10
ABBEYDALE PARK RISE 17	COVERDALE ROAD 7	RIVELIN VALLEY ROAD 6

TALES FROM THE FLYING SQUAD

2016-2018

'It's amazing how much disruption you can cause by stepping out of line a little bit!'

Alice Fairhall, Protester

During the winter of 2016-17, as the protests and direct action grew, a small number of protesters began to be more organised and attempted to cover all of the city, especially areas that didn't have a large or active tree campaign group. The group became known as the 'Flying Squad'. Later, this became a round-the-clock activity as felling crews started working at the dead of night in a dangerous and desperate attempt to fell Sheffield's trees.

Susan Ashworth: The Flying Squad was the best time! Everyone was against us. The arbs, the police, even STAG! That fucking Facebook group full of people making a fuss online and not showing up at fellings! And those *Maintain Canopy Cover* quotes made me laugh my head off! I remember turning up to a felling on Struan Road and some Nether Edgers looking relieved and asking me if I was from STAG. 'Certainly fucking not!' I said.

Paul Brooke: After the Battle of Chippinghouse Road and the police withdrawing from routinely supporting felling attempts, we could save a tree just by getting there and standing under it. By 'just' I mean if you could find them, if you could face the abuse from someone who wanted the tree gone so their car would be cleaner or a 4x4 could park on the verge or some such crap. When you meet people who don't like cherry blossom you really wonder how fucked up things are. Being self-employed I was able to be pretty flexible and respond to messages on WhatsApp or on the STAG

Facebook page. The Flying Squad grew somehow, and gradually you'd meet names in person. I remember meeting Isabel and Angela in Handsworth on an estate where the roads were named after trees. There were local residents who didn't want to lose trees but were terrified of their neighbours. We'd get a message sometimes, other times I'd go and park there before work. The arbs got to know our cars, and once I stopped a felling just by flashing my lights and waving. Then we got smart and decided to let crews get all set up and not stop them until the arb was up the tree – it wasted more time.

Alice Fairhall: We became organised which evolved over time and we became good at quickly responding to call-outs all over the city. The judge even commented in his summing up that at some point we got too good at stopping the felling, and that's when it became a problem. The difference between us and most other campaigners was that the Flying Squad went all over the city, rather than just staying in our neighbourhoods. This meant we got recognised quickly by arbs or goons, but we didn't know what the reception from residents was going to be like – disdain, actively pro-felling or tree lovers. It was people who we knew were willing to actually stand under trees. Even in a big campaign it turned out that not many people, for different reasons, were prepared to do that. People with the willingness to take the risk, the aggravation, the 'who knows what might happen' of doing direct action.

Susan Ashworth: I spent half my time at work on my phone trying to figure out where the crews were and where we were! Angela [Spensley] was fucking hilarious. 'I'm here', she'd text. I'd ask her, 'Where?'. Angela: 'I'm hiding round the corner. Residents are really angry'. 'Angela, love, where are you?'. Then radio silence for a fucking hour! Or me desperately trying to find out where everyone else was, to try and send someone to Angela. 'I'm under the tree'. 'Which tree, darling? Do you need support?'. 'Yes please, they're angry. I'm not making eye contact'. 'Angela, where the fuck are you?!'. Meanwhile everyone who was out and about was saying where they were. Except Paul [Brooke] who would only say once the crew were packing up and he'd been there for ages!

Calvin Payne: It was about being up for it, being available, but for me it was also about getting to wherever we were needed. As the non-driver in the group, I was often after a lift early in the morning. Angela lived nearest, so she often drew the short straw of picking me up first thing in the morning!

Mark: When I started doing NVDA on a regular basis, there were up to twelve separate felling crews active in Sheffield 'smashing and grabbing' as many trees as they could. Teams were being brought in from all over the country to try and overwhelm the protesters. At the time I was self-employed on short-term contracts so I turned down quite a few jobs in order to protest. My income took a major hit, but some things are more important than money. I didn't go to any workshops about NVDA, and I hate to admit it but it was actually Simon Crump's Facebook posts about direct action that first inspired me to start. The first time I did I bumped into Alice on Abbeyfield Road, who demonstrated climbing over barriers and stopping a felling. I picked up the rest as I went along. I found I had a reasonable talent for coordinating protesters as well as taking direct action so I got involved with some call-out groups for local areas and was invited to the 'Flying Squad' group, who would roam city-wide to stop felling, countering the Council smears that we were parochial about our local areas, or only concerned about house prices. I also took exception to this labelling because I had only ever been able to afford renting but loved walking tree-lined streets in my local area. The coordinating was akin to herding cats, as I learned that the type of people who are passionate and bold enough to take direct action against a Council/police/large corporation can often be somewhat highly strung! People flouncing out of groups was a regular thing and certain key protesters refused to be in a group with other key protesters at times, so it all required an amount of diplomacy to make things work.

Simon Crump: I remember in the summer, we all went to Alice's for a very serious meeting which obviously didn't involve any alcohol at all, and Susan [Ashworth] mentioned WhatsApp and that we should start a group so that we could communicate. She asked for suggestions for what to call the group and I thought aloud and I said, ' Oh God!'. And that was it, Susan called the group that! The 'Oh God!' group. That WhatsApp group must have thousands of posts by now, most of them probably just saying, 'Under a tree on xxxx Road, bring food!'.

Susan: I remember that 'meeting' at Alice's! What a motley crew! We set up that group on messenger or WhatsApp. I had to do it for some reason, and when I asked what we should call it and Simon went, 'Oh god'. So that's what it was called. It was still to this day the best online group I've been in. Too fucking funny.

Kaarina: There were halcyon days in the summer of 2017 when Anderson Tree Care would come round and it would be friendly. We'd all be sitting around and Tom Anderson would be there stroking a local cat that had wandered up. We'd be all sweetness and light, saying things like, 'Oh, how did you manage to get here so fast?' and they'd go off again. You remember those days, before the injunction and without the police. It was all just us taking it easy.

Susan: Mainly I covered Millhouses and Ecclesall while Alice went elsewhere, because I worked silly hours and most fellings were stopped early. Millhouses and Ecclesall had no cover at that time. Nobody dared do anything!

Russell Johnson: One incident that said a lot was the Hunter House Road banner, remembering the fallen from the First World War. There's a photograph of the banner of Hunter House Road and the subsequent film of the vindictive arbs [Jason and Dom] – we erected it one night and then the next morning they came and removed it. What was remarkable was that it so irritated the Council to have this banner up, they prioritised sending security as well as arbs. It was on Facebook anyway. They told us we weren't permitted to erect material on Council trees.

Isabel O'Leary: A campaigner from the road was driving along and the Amey vehicle moved in front of him and made him go into the kerb. It was the 19th December [2017] when they took it down, so they put it up on the 18th, and Amey took it down on the 19th.

Russell: It was a concerted effort by residents of the street, and a large proportion of residents at the top end of the street, to express their anger and distress about the Council's intention. There were parking restrictions applied at that time, hence there was an imminent threat to our nine condemned trees. So it was an expression of anger; it was also to gain publicity, of course which it did, and I think the fact that the Council came the very next day says a lot, and the act of making and erecting the banner helped to bring locals together too. It never came back either, the campaigner would have had to appeal to Amey in writing and disclose information, unfortunately.

Susan: I remember the day the police withdrew after Paul Powlesland's Barrister's Opinion got everything paused. There were two fellings at the same time and I was at work, but about to finish. I couldn't decide where to go. Alice was there already, of course, so we went inside the barriers. Calvin and Simon had already been arrested so they were outside the cordon. Finally two police came. Too fucking funny, a tiny woman and a big guy. The little one did all the talking. She most certainly had not been briefed by Stubbs! She said, 'Er, do you realise that it's not safe to stand there?'. Alice said, 'Oh no, isn't it?'. She looks at me. We shrug. 'Yes, you need to stand outside the coned area', tiny cop said. This was at the time when Stubbs was giving us ten minutes to get out from under trees. These two had no idea! Alice said, 'Oh well, can we think about it?'. To which she said, 'Yes, of course'. I asked, 'How long have we got?'. And she said, 'Well, we've got all day'. 'Ha!', Alice said, 'Oh right, great!'. They were shit! Then word came through about the pause after Paul Powlesland's opinion piece and we all went home.

Simon: I remember saying to the coppers, 'You'd better call your boss, new shit has emerged'.

Susan: Yeah, haha! I remember! The Dude in *The Big Lebowski*! I loved how we knew about it before the cops did! I also remember Simon showing the arbs the photo of Stubbs made up as a woman on his phone!

Calvin: One occasion when I was first one on the scene was on Rundle Road in Nether Edge, a two minute walk from home. I found a crew set up around a tree, went inside the small, rather half-hearted, work zone, to be approached by a female arb who I never saw again. She told me that she was an environmentalist, and, as a tree surgeon, a tree lover. I told her that she wasn't a tree surgeon anymore, that maybe she used to be, but this wasn't tree surgery. She was upset, because of what I was saying, but also, I suspect, because it was true.

Isabel O'Leary: When everyone was going through the events on Chippinghouse Road and celebrating the police withdrawal, that weekend I was preparing for elsewhere. I remember saying, 'You're all in the wrong place'. I went to Handsworth and posted flyers through the doors. A resident came and had a go at me, another was moaning to an arb about their issues with the tree outside their house. I met John [Errington] there, I said to him, 'Do you want to do the other side of the road?'. He said, 'I'd rather stick with you, I know what they're like round here!'. So we finished posting on that road, and then I went down to Larch Hill. After John left I met Heather [Russell] and gave her some of these leaflets, posting there and a guy came out, now these leaflets we were giving information about the campaign and what the Council were planning to do and people could join the campaign by looking on Facebook and get more information there. And he came out shouting about, 'These are the wrong trees – they are forest trees and they've planted them in the wrong place. What are you gonna do when that tree falls down on my car?' and all of that nonsense. From our work that day people got in touch with us to get involved, but needed to stay under the radar. They kept a watch on the street from outside at first, but later from indoors as the hassle started.

Alice: There was always a mix of reactions in every area, but sometimes the tree-friendly people would be a bit scared to speak up. Handsworth was like that; we had a 'mole' who would message us as soon as a felling crew arrived. We would often make it over there very soon after they had set all their kit up, much to the disappointment and slight confusion of the arbs who had their own mole telling them if we were there or not. In many areas

I'd get butterflies, especially when there wasn't much back-up and you set off prepared to stop the felling alone. My head would be full of questions on the journey: 'Will I get over the barriers, which arbs will it be, will the residents come out and be aggressive, will they pack up straight away?'. Because there were people in Handsworth who would inform us when the felling crews had arrived [and] the arbs used to ask, 'How did you know we were here?'. Sometimes we got lucky, but there were also plenty of people who tipped us off but were afraid to show their faces because of the attitude of the neighbours. There wasn't much back-up, we were a small group stretched across Sheffield, and we all had to be prepared to do it alone when necessary. In those days I was a bit cocky, I felt invincible.

Calvin: Handsworth really wasn't very pleasant. I went there with several of our regulars, sitting in the café with Alice, or in cars with various Flying Squad members. One day I got there with Simon and we didn't even have to get out of the car to stop the work! Just arrived and sat there. Angela did great work on the Triangle estate too, she was there on her own at least once, standing her ground, which was never easy to do.

Susan: The best Angela story is when we couldn't find her and it turned out she'd gone to one of the worst places alone! I had to talk her through it on the phone. She was terrified, but she stood her ground. It was when the arbs, the police and residents were trying to make out we were intimidating and violent. Instead they got lovely, quietly-spoken Angela not making eye contact. Brilliant! She saved that tree for a day. God, I loved that woman. She was just brilliant!

Isabel: I went to Willow Drive in Handsworth after work one day to find they were trying to fell an Italian alder that I'd been trying to protect for a long time. It was a beautiful tree, lovely, with absolutely nothing wrong with it or the pavement around it. Phil [Yates] was there and one of the arbs was stopping him from getting inside the barriers; they were moving sideways against each other. I saw them and just went and hopped in at the other end! The arb turned away from Phil and just said, 'Oh'. Darren Butt was there and I asked why the tree was being felled, and he didn't know.

A resident was shouting at us, but eventually the crew gave up and Butt helped them pack away the work zone which I have a photo of! He'd taken a picture of me on the same road that was used in the injunction case. The next day I was working later on and waiting by the same tree. I had a long chat with the arbs about the Amey contract. When I was needing to leave I started putting out call-out messages. No-one had arrived so I tried to back off quietly, hoping the arbs couldn't see me from where they were sitting in the van. As I was leaving I saw them go up the tree so fast that one of their hard hats fell off. I drove away and haven't been back since, it would be too sad.

Russell: One of the arbs asked me once, 'How do you lot always get here so fast?', to which I told him, 'We've got eyes and ears everywhere in this city!'. In places where there is a negative attitude to the campaign the Flying Squad had the advantage of not living there. We could come and go; sometimes that was a good thing. In Handsworth I was told, 'I'm going to fucking shit on your head!'. I enjoyed being polite in response to that sort of thing!

Susan: I remember an arb looking down and saying, 'When the fuck did you get there?'. Like a ninja!

Isabel: On my way to work one day in May 2017, I attended a call-out on Abbeyfield Road in Burngreave and stepped over the barriers to stop the felling. I kept looking at my watch thinking, 'I've got a patient in twenty minutes!'. An Acorn vehicle came and they were taking pictures of me. I now know this was the beginning of evidence gathering for the injunction. I had to go and the first back up that arrived took my place inside the work zone. I saw a campaigner standing with her baby outside the barriers when an Amey supervisor went bonkers at her: 'You're an unfit parent! You're endangering that baby!'. I said to him, 'What on earth are you talking about?'. Next thing I know he was on the phone and soon afterwards the police attended. She was subsequently reported to social services.

Susan: One of my favourite moments came from a day in Millhouses. I spent half my time looking for arbs. I found a crew on Whirlowdale Crescent (again) and stood under the tree. They didn't bother setting up properly, just sat around chatting and threatening me with that piece of paper they had at that time and hoping I'd go. A woman came over and asked what I was doing. My heart sank because you never knew which side they'd be on. The arb was looking smug because she really looked like she might have a go. But she listened to what I told her and she said, 'I'll stand with you'. Me and the arb just grinned at each other and I told her not to accept the paper and that the police might turn up, etc. Eventually the arbs went and I went off looking for them again. I never met the woman again, but a few days later I saw a picture of her on the STAG Facebook page standing under the tree outside her house. She saved the tree that day. I spoke to Angela online that evening, she'd rocked up to the felling, and she said she'd started advising the woman on what to do, but she said, 'It's OK thanks, I know what to do'. Yay! I think a lot of that stuff happened.

Calvin: Abbeyfield Road was a regular trip for us. Not far out of the city centre, with a small local tree group, it's a nice road with mixed feelings amongst the residents about the trees. One resident told Darren Butt that he could get some friends together to drag me and Simon into the park and sort us out, to which Butt replied, 'I can't let you do that, I'm afraid', in a manner that suggested the idea had some appeal! One tense morning on the road Paul [Brooke] started talking to an anti-protester resident who was doing some construction work on his house. The next thing we knew Paul was pushing a wheelbarrow back and forth helping the guy! Later the same bloke came out with a pot of tea, looked at the group, then at me and Simon, and said, 'But not those two'.

Heather Dewick: Initially, local support against the felling of Abbeyfield Road trees was promising, but sadly very few people actually participated in the protests or even supported us in other ways. I was told, by a local climate change campaigner, after we had yellow-ribboned threatened trees, that these mature, pollution-busting trees were a 'nuisance'. The irony. In the end, as local support proved so poor, we were pleased to accept help from more experienced (and more organised) protesters from across the city.

Calvin: Abbeyfield Road was also where I met, and stood under trees with, another sadly missed protester, Mary Marshall. She was a real character from the protests and one of the first to have a Facebook pseudonym, 'Cherry Picker'. Later on, in the injunction days, lots of people came up with tree-related alter-egos!

Heather: We were on the receiving end of some nasty aggression from a few local men, one of whom took to spitting at me as I stood behind barriers. He was incensed by leaves dropping on his car. He continued to harass me for weeks afterwards, even as I walked to the bus stop. This frightened me and put me off standing on the street alone. I took on a different role using a dog walk as an opportunity to patrol and spot incoming arb and Amey vehicles, and then alert braver people than myself. I did get behind barriers on occasion though. One notable day, I was standing with Simon when Darren Butt arrived. He was so narked at our presence that he lost his temper with Simon, pushing the barrier at him. Later, we were joined by a local Labour ex-councillor, who'd come to have a go at us. He actually used the words, 'Go back to where you came from', a loaded, offensive phrase generally, even more so when uttered in the most ethnically diverse area of Sheffield. And these people are meant to represent us.

Isabel: Two days after Russell and I got married we were out standing under trees. I told an arb that and he asked, 'What does your husband think?', to which I replied, 'He's under a tree further down the road!'.

Paul: Thorpe House and the Norton area was a long drawn-out battle and Amey brought in crews from all over the place and tried these mass fellings. 'Treemageddon' was an onslaught – they were pitching up on a road with three or four felling crews and support. Once, on Thorpe House Rise I think, I spotted a crew when doing a quick scout round before work. I parked and thought I'd wait for the crew to set up – I messaged out that there was a crew. As I was doing that another crew arrived so I rushed to the first tree and stood under it. I was frantically getting messages out and jumping between two work zones as more crews arrived. You never knew if or when others would turn up but you hoped. I think Simon pulled up first and I'm relieved and I was shouting that there are more crews 200 yards up the road. Within minutes I'm told that all four crews have been stopped. Heather was there, and other people I didn't know, and just a few branches lost. I think Russell took over 'my' tree so that I could go to work. It was full-on but I felt part of something that seemed important.

Simon Crump: On 24th May 2017 I finally understood why people like to watch sports so much. I got to Thorpe House Rise early in the morning, and there were five crews there with a potential eighteen trees to take out. Paul was already there, and was flip-flopping between two cordoned-off areas. I was there next, because I'd had to spend a while in make-up, and so I got myself under a tree too. I was feeling a bit shaky about the whole thing, having had an altercation on Abbeyfield Road the previous day with Darren Butt in the morning, and then been knocked out cold by assorted angry civilians on St Ronan's Road in the afternoon. Then, Patrick [Beaumont] turned up. He made a run for a barriered-off tree pursued by an Amey worker who was trying to body-block him. And Patrick made it, and he got himself into the compound up the road from Paul and me. And I swear, the last few seconds of that run were in slow motion and the music from *Chariots of Fire* was playing. It was fucking brilliant!

Russell: The Flying Squad, as we were referred to, was about covering the whole city where needed. We always encouraged residents, of course, to defend their own trees, but where there was hostility, notably Handsworth and the Thorpe House area, we had the advantage that there were some of us who didn't live there. The advantage is in being able to leave at the end of the day's work, although anywhere we attended was because we had a call-out, because there was someone on the road who cared.

Calvin: Going to other areas of the city always caused a dilemma for protesters. In practical terms it was hard going to an area, or a street, without knowing anybody, without having much information on the trees or what the attitude of the residents was. The unpredictable nature of direct action was difficult enough without throwing more issues into the mix! Then there was the moral question, should we get involved on streets we didn't live near or were even going to visit again? When we spread ourselves across the city we were accused of being 'professional protesters' or were being 'bussed in', but when on our own patches we were called 'nimbys', or only concerned with 'posh' areas of the city. We really couldn't win, and I think we tried to apply the same approach wherever we went, that every area of Sheffield was as important as everywhere else.

Martin Pickles: I saw an Amey or Council worker photographing me on Abbeyfield Road. I got a bit narky, I didn't know why they were doing it. We got a call about Shirecliffe Road nearby. I arrived and there was no-one else there yet. I really felt in the thick of it then. I texted and phoned Graham [Turnbull] asking, 'What should I do?'. 'Whatever you feel comfortable with, whatever you think is right', he replied. The arbs' truck made up part of the work zone and I looked at the space underneath. One of the arbs saw me and said, 'Don't be stupid'. He was right so I chose my moment and scrambled over the barrier into the work zone. I had no idea what was coming with the injunction. In the evidence, they had one of my Facebook posts from that day saying, 'I STOPPED THEM!'.

Simon: Abbeyfield Road was the first time they used the 'photo arbs', men who dressed as arbs but were there purely to take our pictures, which we later discovered was because they were preparing the injunction and assembling evidence of our supposed wickedness and the 'damage' we were causing, which was bullshit, obviously.

Heather: Amey used the pedestrianised part of Osgathorpe Road to store hundreds of barriers. This made the street hard to navigate, and even harder when most of them fell over. I tweeted Amey. Their response was to send someone out to put yet more barriers around the fallen ones. This sounds funny and it is, in typifying rank incompetence, but it summed up the disrespect shown towards the area generally.

Simon: I encountered an extremely unpleasant man under a tree on Thorpe House Rise, name of Mick. My impression was that he fancied himself as the local tough guy. He had a right go at me, got right up in my face and said, 'I'll find out where out where you live one day, it might take ten years, but I'll fucking find you'. Which all seemed a little bit too biblical to me. 'I'll save you the bother', I said, 'You can have my address now if you like, and I've just recorded you threatening me on my phone'. Then I ruined it all by adding, 'You fucking wanker'. Paul took me aside after that, in an attempt to impart a few basic people skills, which as it transpired he had a shedload of and, as Paul observed, I did not. He said that I must have been at the back of the queue the day they'd handed those particular skills out. 'When an angry resident confronts you Simon, a) don't respond to their threats, and, b) don't try to be clever, and c), do not, do not ever call them a *fucking wanker*, but do, d) always respond instead by saying something nice to them, it can often take the heat out of a situation', Paul said. So next time I was confronted by Mr Hard Man, I decided to adopt exactly this tactic. Mr Hard Man got inside the barriers around the tree with me and forced me into a corner, invading my personal space and rather hurting my feelings in the process. I could tell that he was angry and in no mood for a friendly exchange of banter or witty repartee, but even more disturbingly he had his top off, revealing a sizable pair of man boobs. 'Good morning, Mick', I smiled, 'Nice tits'. It didn't end well. But the bruises did fade in time. And scars add character after all, don't they?

Calvin: I spent most of a day under the same tree on Abbeyfield Road with Simon. Darren Butt came over to us when the arbs started packing up and said, 'Well, I've wasted your day', which I think fundamentally misunderstood what we were about! We were only ever 'playing for the draw'. We just wanted to finish the day with the same number of trees as we started with. John [Errington] from the Burngreave tree campaign group came over to see if he could get us anything. He went to fetch some food and drink for Simon and me, and kept us company for a while. John was terminally ill with cancer, but still came to check on us and help out. He is much missed amongst many people in Sheffield, well beyond the tree campaign.

Simon: John was a top bloke! That day on Abbeyfield I'd given him some money to get crisps and sarnies for Calvin and me. He looked so ill. John was definitely one of the good guys and I miss him.

Alice: Our WhatsApp call-out group was good fun, lots of chat and stories. Kaarina [Hollo] used to make me laugh the way she told it, 'They are having a go at me for getting the bus one minute, then asking how rich I am to not need to work the next!'. Looking back now there were a lot of not very insulting insults really.

Simon: I turned up early one very cold morning on the Triangle estate in Handsworth just as a felling crew were setting up under a tree, and on this particular occasion it was just me, all on my tod. I'd recently discovered Joules Tweed coats on eBay and I was wearing my latest, loudest purchase. I parked up and then got inside the barriers. And an arb said, 'It's the cunt in tweed'. 'That's fucking Doctor Cunt to you, pal', I said, 'in fucking tweed'.

Calvin: The arbs, bouncers and barrier men spent some months trying to push my buttons. I'm pretty transparent, really, so I'm surprised it took so long to manage it. They thought I was insecure about my dress sense and charity-shop clothes, which was very strange because I dress myself in the mornings! What did they think was going to happen? That I was suddenly going to be embarrassed about scruffy trainers and pink polo shirts? Then they started saying that I was some sort of sidekick to Simon, which was also not very insulting, and by the way hopelessly wide of the mark. Finally they crossed a line and were very personal. I spent months having the physical signs of a health condition loudly pointed out on the streets. That was a line I wouldn't cross with someone on the other side, despite all the things I dished out to them during the protests. I could have taken that issue further and I hope they know now that it was wrong.

Simon: It was the day after the first pre-injunction hearing at Leeds High Court, when SCC had drafted in extra felling crews from all over the country in the expectation of obtaining an interim injunction (which they didn't get). Calvin and I had been following the same out-of-town crew all morning in the car, an aged Clio. Every time they stopped at a tree, we waited for them to set up and then got inside the barriers with them, and I thought that it would be funny to say exactly the same thing to them every time we stopped them from working, as if it were always the first time we'd met. Calvin didn't think it would be funny, incidentally, and refused to participate. Which, I have to say, is just typical Calvin. Anyhow, we'd get inside the barriers, I'd say, 'Hello, what a lovely morning, how very nice to see you', and then offer the same arb exactly the same colour wine gum every time. Travelling in convoy to the next tree, they'd already tried to shake us off a few times by indicating one way and then turning the other, which had resulted in hasty three-point turns and also some of the world's slowest car chases, and by dinner time we'd parked up next to them in Morrison's Woodseats car park to keep an eye on them. They didn't seem too pleased about that either. An arb got out of the truck and came over to the car. It was a lovely, sunny day and we had the windows down and the key turned in the ignition, so the fan would work. The arb leaned into the car: 'Not got much juice left, have you?', he said, which was true, we'd been running on empty all morning, and with no time in between attempted fellings to fill up. As the felling crew set off after dinner, it quickly became evident that they had 'a cunning plan', and they headed out into Derbyshire with us in not so hot pursuit. Their 'plan' was to run us completely out of petrol, leave us stranded, and then head back into Sheffield and get some trees down. After a bit, we had to turn back. It was a simple case of cutting our losses and getting back to civilisation, or facing the horrible prospect of living out the rest of our miserable lives wearing animal skins in a field in Derbyshire, eating roots and berries and worshipping rocks that looked a bit like breadcakes or something. So at the next 'T' junction, when they turned right, we made a crucial lifestyle choice and turned left, and there was much jeering and some rather crude and hurtful hand-gesturing, on their part, as a result. We made it on fumes back into Sheffield, got a few quids' worth of petrol in at Bramall Lane Services, and then, just on the off-chance, we drove back up to the first tree we'd stopped them having that same morning. And there they were, just setting up. So we got inside the barriers, Calvin and me, just like we always did. 'Hello', I said, 'what a lovely afternoon, how very nice to see you', and then I offered the same fucking twat arb exactly the same colour wine gum. You should have seen the looks on their little faces.

Paul: Simon and Calvin were hilarious, infuriating and determined in equal measure! Their commitment to resisting by direct action inspired me but there would always be drama when Simon was around – something would always kick off! The arbs really hated them, I reckon, and it did get personal. STAG was the campaign, but it was all about individuals taking a stance all across the city, wherever we could get to in time, that prevented hundreds of healthy trees being felled. It meant there were posts on the STAG Facebook page to celebrate rather than simply pictures of angry people next to felled trees – there had been enough of that. As a 'Flying Squad' we did what we did and no one was in charge. For Simon and Calvin I don't think it could have been any other way!

Mark: When the Council started handing out 'pre-action protocol' notices threatening people with court action unless they signed a draconian undertaking (including not posting encouraging comments to fellow protesters on Facebook) I realised it was time to go undercover to continue protesting effectively and not risk incurring huge legal costs. So the pseudonym 'Theresa Green' was born for my posts on Facebook, and I didn't tell anyone my name during NVDA. The pre-action protocol was quickly followed by a court summons for the group of named significant protesters; this included most of the 'Flying Squad', who then became fully occupied preparing a defence for the injunction hearing. I'd slipped under the radar so far, so mustered others in the same position into a new group to continue where the Flying Squad left off and drummed up support through Facebook. We feared that the Council and Amey would use the court dates to maximise felling attempts, and thankfully we managed to keep them at bay. After stopping a felling in Sheffield on the second day of the hearing, I got a text message from one of the defendants in court with a quote from Paul Billington's witness statement to the trial: 'Between 1-29 June 2017, 472 fellings were attempted, 329 of which were abandoned due to direct action'. I was delighted!

Alice: One day in Fir Vale, I was on my own attempting to stop two felling crews, when a group of young men came out of a house across the road, pointing fingers at me like they were pointing guns. They shouted at me for a while, then one shouted, 'You're gonna get raped'. It was probably the first time I felt genuinely vulnerable and acutely aware I was the only female in what felt like a sea of men. For the only time in the eighteen months or so of active protest, the arbs were quietly protective of me. They had seen me shouted at countless times, but it was like we all knew it was different that day. A few minutes later, one arb sat down near me as I tried to make myself invisible to the residents behind a chipper. Another arb later was sweeping near me and said very quietly, 'Don't worry, we won't let them do anything'. Deep down they didn't want to see any of us hurt; we might not have been friends, but we were civil and peaceful. I put out a call-out message saying, 'I really do need help here, I think I'm in trouble'. I couldn't have been more relieved when Paul turned up. He'd come away from work when I messaged and realised I must be in real trouble. The evidence gatherers still used photos they took of me that day in the injunction case, though!

Susan: That tree on Whirlowdale Crescent that the ITP supported saving: I think it was Dronfield Tree Services helping Amey. I turned up on my rounds. Alice was watching from her window. I didn't know until just after. They nearly squished me and anyone else who might have turned the corner and walked under. It was shocking and so fast. I ran to the tree and a guy said don't go under. Which of course made no difference, but he stopped me passing. He said they had started. But I knew it can take ages to fell a tree, so I wasn't bothered. I ran round the other side to get under, but some stupid woman started having a go at me. I started filming and then the whole fucking tree came straight down. No way! Fucking hell. I was so shocked. We'd not seen that before. I remember being in Alice's house after, and Justin [Buxton] turned up. He was almost in tears. That was his tree, if you know what I mean. We were all pacing around. Like, how the fuck can we defend trees from this? It was crazy quick. In the end I left and a couple of minutes later got a call from Alice saying stuff was happening on Silver Hill Road not far off. So I stopped off there. A different chipper from the one I'd just left. The Dronfield's guy was shouting at a resident in the street. I had to get in between them and calm it down. It was relentless, and most of it never got on STAG. He apologised to the guy, and it was just another intense moment in a fucking never ending nightmare.

Mark: In October 2017 there was a narrative from the Council and Amey about 'intimidating' or 'violent' protesters. This involved a promotional video with interviews with a few 'distressed' residents, who couldn't sleep for worrying about the people standing under a street tree outside their house, and a shoddy piece of client journalism in the *Sheffield Telegraph* about the 'unacceptable face of protesting'. All this was leading up to the mobilisation of teams of nightclub bouncers, SIAs, to 'protect' the felling crews. Actually it was to remove the protesters using 'reasonable force'. The 'violent and/or intimidating' narrative had been fed to the bouncers before they set to work pushing us about. When they got to see that we were in fact peaceful, reasonable people, they were clearly perturbed, but not for long! I'd been thinking that we could help debunk this 'intimidating' narrative if I 'bunnied' without a mask. Unfortunately, the day I chose to do this was the first day the bouncers appeared on our streets. 'Mr Uppercut' wrestled me to the floor in the first safety zone I entered on Coverdale Road, then I was again pushed over in a zone on Crescent Road by more bouncers that afternoon. My efforts to appear 'non-intimidating' became something of a moot point after footage of the bouncers dragging a different bunny about on Lismore Road went viral that night!

Roger Doonan: I used to bait 'Mr Uppercut' on purpose, really used to push him in the hope that he would push back at me, but he never did. I really wanted him to, and I was very vocal with him, but he never rose to it!

Mark: Clearly the Council and Amey were becoming ever more desperate by December 2017, so some bright spark decided that they would start sneaking in at night to remove the oversailing branches with hand saws, making the rest of the tree easier to dismantle by the crews in daylight. This was clearly against all Health and Safety rules, they had inadequate lighting, and were working over parked cars, but when did they worry about things like that!

Alice: Mark realised we needed a rota in response to the night raids. It covered midnight to 7am. We'd do a four-hour stint, go back to bed, then be up for work. It was the most nuts time of my life; to be at work the next day living a normal life was an unusual experience to say the least. Crazy times!

The outcome has been positive, and the experience was positive too. It's amazing what disruption you can cause by stepping out of line a little bit!

Mark: We realised we had to try and stop this latest tactic, so a group was formed of insomniacs, night owls, committed protesters and some generally unhinged people to patrol the city on the lookout for arborists. If interrupting felling by day was intimidating at that point, the prospect of disturbing a crew at night who were already pretty rogue, when you were alone or in a pair, with limited ability to film any conflict, was another level again. Finding volunteers was consequently pretty difficult, so we played the Council at their own game by this point with our own propaganda. 'Theresa' posted on Facebook about the teams of night patrollers all over the city, when in reality it might have only been a couple of cars doing a circuit of the city at two in the morning, while another guy cycled around Nether Edge!

Alice: I couldn't have kept it up much longer but then nor could the Council – thankfully. The outcome has been positive, and the experience was positive too. It's amazing what disruption you can cause by stepping out of line a little bit!

Roger: On night patrols I'd drive round wearing my hi-vis that I use at work. On a couple of occasions I saw another campaigner driving around and we saw each other checking each other out. I'm sure I was mistaken for one of the other side for a while. Technology was central to our organising, especially with the 'Night Owl' patrols. We relied on modern technology so much, which would have been so different not so many years ago. I went through the process of moving from Facebook messages to being on the closed WhatsApp group, where people had to vouch for you, where you had to prove yourself. It was fun and good-spirited despite everything we went through, but there was understandable anxiety about infiltration, with new people being screened.

Mark: We carried out the patrols until Christmas 2017 but began to question if we could really keep going in the New Year. People were patrolling by night then bunnying the next day and getting burnt out in the process. We hoped the Facebook propaganda, plus actually catching them up in the act a few times, might be beginning to make them question the tactic, but it all came to a head one night on Rivelin Valley Road in January 2018. We'd managed to block their vehicles into a parking bay when they'd turned up to try a felling, so there was a stand-off for a couple of hours. One hot-headed arb was clearly psyched up by a chat with their field manager and charged up a tree despite two protesters standing below it, then came down and started scuffling with protesters. Police were called [by the protesters] and we encountered the usual bias where they talked to the arbs first and assumed we were at fault. Eventually, with the help of Jaq [Pixelwitch]'s video footage, we made it clear that we were the ones making the complaint, along with a complaint made about the same field manager intimidating and verbally abusing a lone female protester earlier in the night.

Simon: I remember that on the 'Oh God' WhatsApp group one day Paul had put a couple of diagrams up about geckoing which started a discussion about the tactic. Since then pretty much everyone I've talked to has said it was Justin [Buxton] who perfected geckoing into the extremely effective tactic it became. We'd have been fucked without it on so many occasions!

Calvin: When the call-out came for the war memorial trees on Heathfield Road in Frecheville it was out of the blue. With memorial trees under threat being news on Western Road in particular, we had become aware of other streets in Sheffield with trees planted after both [world] wars in memory of the men who didn't come back, but this was the first such attempt to fell them. It was Bryan Lodge's council ward and I feared a set-up. Would there be residents lying in wait for us? As it turned out there was a lot of support from locals, and a lot of concern that there had been such little notice of the work. One of the trees was hollow when felled and the arb concerned took great delight in standing inside the stump to illustrate the fact. Whether this meant the tree needed felling or not I don't know, but it also led to the bizarre spectacle of Bryan Lodge parading a slice of the trunk around the following week's full Council meeting. I honestly thought someone had photo-shopped that image of him when I first saw it.

Martin: At the time I'd been finding out more about my relatives and their role in the War, things that had never really been talked about in the family. I'd been to a meeting on Western Road [in Crookes] wanting to know what we as a campaign were going to do about the memorial tree issue. I said to Councillor Lewis Dagnall [then Sheffield Council cabinet member], 'How are you going to do it? You know there will be hundreds of us here'. I went to Frecheville when I saw the call-out. Well, I lost it a bit! I made a bit of a spectacle of myself. I got the impression we had been duped, that it was a trap. Someone put a film of me on Facebook that night and I asked them to take it down. I was angry and upset when I saw the felling, but I felt a bit of a fool afterwards.

Alice: As well as feeling I was helping the cause, I loved being part of the Flying Squad, we were like the inner team, we got on and looked out for one another. It was a positive experience. We learned from the not-so-good things that happened. Residents shouting at us to get a job or go home was very minor stuff and I soon realised that when we were being genuinely threatened. I became a better campaigner for all of the experiences, including the upsetting ones.

Simon: Above all it was fun. It's true that we got knocked about and abused, and it's also true that other parts of our lives suffered from all the time we devoted to stopping the fellings, but it was still the best fun you can have with your clothes on! I was very well-behaved at school and I was never in a gang or anything like that, and so I like to think that my time as a member of this small gang of reprobates who other people called 'The Flying Squad' (we never called ourselves that) made up for my rather dull childhood in some way. I didn't even know how much of a show-off I really was until I got involved in all this stuff. I was alive, I was in the papers, I was on the radio and I was on the telly. It was fucking great!

Alice: One of my favourite memories was when someone had posted on a local Facebook group a photo she had taken from her window complaining, 'One person holding Amey up from their jobs and it's still going on'. She was having a go, but I thought to myself, 'Oh, that's me! Brilliant! Have I really been here for three hours?!'.

Susan: I have to say I loved it when other people got involved. Like 'Running Man' and the others, but I really loved our 'them and us' attitude when we were against it. Just us few. We spent all that time trying to get people to join us, and then somehow it wasn't the same. But hey, we won. And nothing stays the same. And I fucking love trees!

Alice: Some great campaigning work was done behind the scenes. The questions, petitions, FOIs, lobbying, meetings: so much by so many people. I couldn't have done that, I was very much a foot soldier who held up the felling, held it up so others could do their stuff. I sometimes feel frustrated that some in the wider campaign didn't really know what the foot soldiers did though.

Susan: Also the threats hanging over us all. The threat of arrest, of jail, of fines, of violence, of having our houses taken off us. Constant threats. I was scared. Amey was a big company and big money was involved. But the camaraderie of the Flying Squad got me through. Even if half the time someone wasn't talking to someone else. Haha! The tension was incredible. And there was no stopping us! I fucking loved all of it! I felt alive. I had a purpose. Totally knackered in every way. A lot of people were twats, on both sides. But we were right and it felt good. I remember one weary Sheffield arb saying, 'You can't beat the Council'. I couldn't stop laughing at that!

THE REAL 'PERSONS UNKNOWN'

'What I would really like is for as many people as possible to break the injunction on Monday morning.'

Calvin Payne, Protester

In August 2017 Sheffield City Council secured their High Court injunction against peaceful protest by tree protectors. That injunction was awarded against both named campaigners and also what are legally known as 'Persons Unknown', that is, unnamed people who may, at a later date, breach the terms of the injunction. This led to the appearance of people who disguised their identity when carrying out peaceful protest and gave rise to the term 'bunnies', which became an integral part of the campaign, as protesters found new ways to fight back against the tree-felling programme.

This chapter consists of the accounts of some remarkable people whose identities are still secret to this day. Brave and determined, but also clever and inventive, they defied and defeated the Council, the contract, the police and the courts. In doing so, they often adopted different personas, different walks, stances and even different voices. In short, they had to become someone else. It was extraordinary. This is their story in their own words.

Calvin Payne: It took five weeks between the injunction being awarded and the first breach that I am aware of, which was on Dunkeld Road. I had been talking about breaching the injunction, but hadn't done it despite having plenty of opportunity. That was a rough time, watching fellings that we would have prevented a couple of months earlier. I didn't really have a plan, but when others were inside the barriers I was relieved to join them. The decision was made for me. I regret not breaching the injunction earlier, but also that I took myself out of the game by getting my suspended sentence when I did. I was frustrated standing and watching felling, then starting going inside the barriers again, then back to watching helplessly. One thing I knew was that *somebody* had to face it down and show it could be done. One of my convictions was for encouraging others to breach the injunction (including using the words at the top of this page on social media), so I'm just so pleased that many others followed. Many braver and cleverer people than me – climbing over fences, getting roughed up, punched and kicked, and threatened. I'm proud to know them.

'JESSICA RABBIT'

Somewhere along the way I worked out what I would wear if I was going to do the same as others were doing. I got together some old clothes and a hat and coat that I hadn't been seen wearing before. It was a gradual thing – am I going to do this or not? If I had been caught and convicted of breaching the injunction, I would have lost my professional registration. I would have lost my job.

I had a day off work and decided that if a tree that day needed a bunny to protect it, wherever it was, then I would try and do it. I changed glasses, rings, put on gloves, a hat, and headed out. They went for 'Milly' [on Millhouses Lane]. I left my car nearby and walked to the tree in my 'bunnying' gear, by which time the felling crew had been seen off. Then I got a lift to Abbeydale Park Rise, then on to a street near Ecclesall Road, and finally a call-out on Struan Road in Carterknowle, where I finally got my chance.

I was just following the action. If it needs doing then I'll do it. On Struan Road there were only three or four protesters and a lot of barrier crew and arbs. There was a small gap under one side of the fencing and we created a bit of a distraction on the opposite side of the work zone whilst two people lifted the fence a little bit for me to get under. Having got in the work zone I was lying on my back on the ground. The arb team included Jason Wignell; I can't stand the man and he's standing over me taking photos, and he said, 'It's ****, I can tell by the eyebrows'. Trying not to look like he got it right, I managed to get to my feet and tried to keep my distance from him.

The arb crew phoned the police saying that they felt intimidated because they couldn't see my face, and that they didn't know what I was going to do. They had obviously told the police who they thought I was, and the police officer who attended asked me if I was who the Amey crew thought I was. I answered truthfully and said I was. They didn't take any further action and just left.

The crew started packing up and someone came to drive me away. People on our side knew who I was. As we drove away Jason Wignell was in his truck and he began to follow us. We did a few laps of the area to shake them off, but no luck. We ended up driving to Woodseats Police Station to lose them, otherwise I'm sure they would have followed me home. I was taken to a campaigner's home where I changed again and got in a different car.

So I had broken the injunction. I only did it once and that was the day. I haven't even told my husband that!

By that point other people had done it already, and had basically been OK. No-one had been found guilty of breaching the injunction who had been in disguise. I don't know if Wignell was just guessing it was me. On other occasions they wrongly thought other bunnies were me also.

I wasn't on the streets much due to work, which is why I took my chance when it came. I was getting flak from some people for not being more involved on the ground. Part of me just wanted to be able to stand up in future and say, 'I did that'. It was the only thing that was succeeding at the time. I just felt I had to do it. I was petrified, though, and took a risk in a few ways. That probably contributed to me not doing it again too! Even though I didn't save the tree, the more time they spent chasing after us the better. Sometimes it was a case of 'delay, delay, delay', if that was all we could do. I'm really pleased I can say that I did it. One day we will all be able to stand up and say, 'That was us, here we are', and that we are proud of what we did.

'RUNNING MAN'

Something needed doing, I saw what that was, so I did it. Perhaps all those insomniac nights idly watching *Battleplan* and the like weren't a waste after all? At some point hence I may well reflect on the affair but for now I'm merely happy to allow an occasional Cosmic chuckle that perhaps *the* moments I'm best known for are the ones where I'm most unknown; play that harmonica, son, you're just an alias – I came, I saw, I hopped.

'SHAMEY COWBOY'

I was working up the road from Meersbrook Park one day and heard the chainsaws. I came out, still holding a cup of tea, saw a commotion lower down the road at another tree, pulled my hat over my eyes and went over the barriers. I sat by the tree and my heart started racing. I was expecting to be manhandled out of there when a voice I now know was Paul [Brooke] shouted out 'You can't touch him', so I thought 'OK' and I sat it out until the crews went away.

I got this sense of excitement; you don't know what's going to happen. It was a drama, there was a buzz to it. When the Heras barriers started appearing and the SIA guards were used, it seemed as if the gloves had come off. They had given themselves permission to do what they liked to the streets and the residents.

When in disguise I started talking to people whilst holding my nose, which annoyed the SIAs! We started the slow-walk protests on Meersbrook Park Road to delay them setting off to other fellings. We got good at that. The police allowed us twenty minutes but we stretched that out well. One morning, following a night raid, they had left a work zone full of branches and brush. I sat on it like it was a nest and wouldn't let them clear up. I was in my bunnying gear and I just stayed put!

The first time I went elsewhere was to Dore in the winter. I managed to get in the barriers there too, but got a few cuts and bruises and had my glasses broken. I was dealt with a bit roughly then – things had changed in a short amount of time since I had started getting involved. The SIA guards used to shout, 'Get his mask off!'. They uncovered me after a bit of a struggle and I think they were surprised to see a grey beard on me. They thought I was younger! They pulled me about, but I was holding the Heras barriers, and they nearly all came down as they pulled and pushed me! The next day I spoke to one of the SIAs at a different site whilst in another outfit and holding my nose. I asked him about the scrap he had with a protester the previous day and he went on about how this protester was 'a big bloke, really rough!'. I tried to shrink a couple of inches as he was talking.

I adopted the 'Shamey Cowboy' persona in reference to them and their work practices. By then I had also started climbing trees; that worked in terms of stopping work and also got me out of the way of the SIAs. As I did more and more, I tried to throw them off the scent. One day, I got someone else to wear the cowboy outfit. The SIAs looked at his feet and straight away they knew he wasn't me. They called him by his name, so that didn't work!

I wasn't worried about court or prison. I've been inside before, including for assaulting a police officer in my youth. That's not who I am these days and it was a long time ago. Ironically, Borstal taught me climbing as they were hot on training and fitness, and those skills were put into practice during the tree protests!

I was arrested twice during the protests. Once I was de-arrested, but on the other occasion I refused to give my name and was taken to the detention suite at Shepcote Lane in my cowboy outfit and mask. When we got there, I made them lift and carry me out of the van. I was only held there for three or four hours. I think they had looked up my record and were worried about my previous!

I missed the final day on Meersbrook Park Road as I was at a funeral. I would have had fun if I'd been there. It did look good on the footage. A great day! I pass the tree I saved every day. I give it a hug. It knows me, I believe that.

What was great about being a bunny was the network and support. Food and drink, clothes, things to sit on, something to read – it was great. People who didn't, or couldn't, take part directly in the action helped with that part so much. There was a role for anyone and everyone. We've inherited a tree community, a tree family. That is the beautiful legacy. Not only did we succeed and win, we created a family that will endure for a long time.

'DURACELL BUNNY'

The first time I 'bunnied' was on Meersbrook Park Road, sitting in a deckchair with another protester. I just delicately lifted up the plastic barrier and went into the zone. A bit different from the later events! It was a very civilised affair. Yes, the 'butterflies' were in my stomach fluttering away, my breathing was more rapid than normal, but the conviction to halt, or at least delay, the unnecessary felling of a beautiful healthy lime tree on Meersbrook Park Road was foremost in my mind.

At the beginning of January 2018, I made a decision to take a week's annual leave from work. I knew I needed to make myself available, as across the city things had ramped up significantly and I wanted to ensure that I could be available to respond quickly to call-outs. During this week I was particularly active: on one day I 'bunnied' at MPR, onto Spring Hill Road, down to Rivelin Valley Road, back to Spring Hill Road, and then back to MPR. The crews were certainly giving us the run around and it was that day that I became known as the 'Duracell Bunny'.

Bunnying at these locations was relatively easy, as the barriers used were the low plastic barriers, which were fairly easy to get over, but at times there was some escalated behaviours from the barrier guys who would physically block, push, shove and grab my arms, causing bruising, but with a scenario of playing 'cat and mouse' I was able to out-manoeuvre them and clamber or fall over the barriers. However, then came the Heras fencing, which are galvanised steel mesh barriers that stand over six feet tall, and this was a whole new ball game which made life difficult for the bunnies indeed. Also during this time, Acorn had been out during the night carrying out the ghastly action of butchering our trees, by removing branches that over-sailed parks and public spaces. They had chosen to do this as they knew at that time campaigners were able to safely stand under trees in a park or on private property (with permission), without risk of breaking the injunction. There was a group of us who attempted as best we could to patrol areas most under threat each hour from eleven at night through to the following morning. There were only a small number of us, but we did the best we could.

The day that Simon [Crump] was arrested, we had reached an impasse on Meersbrook Park Road and had been peacefully carrying out a slow walk to stop the crews and vehicles from moving on to other felling sites. As we had plenty of bunnies, and it appeared that Acorn had had enough, I left and made my way down to Rivelin Valley Road where there was a crew setting up opposite the fire station. I was able to get inside the zone, despite two barrier guys attempting to obstruct me from accessing the site. I experienced some rough handling, but I had committed myself and was not going to be intimidated. We played a game of 'cat and mouse' for a few seconds, and I was able to breach the barriers. This was a long protracted 'sit-in'. I found it quite a lonely experience as there was some distance between myself and the few supporters who were across the road. I was also under close surveillance with the evidence gatherer nicknamed 'Boris' and his video camera filming me close up. I kept my head down and tried not to think about the fact that my bladder was crying out to be emptied.

It was all worth it, as again the crew gave up and I was ferried back to my car after a lengthy period, taking care to ensure that we were not being followed. On 11th January 2018 I had been woken up by a call at four o'clock that morning to say that Acorn arbs were butchering a tree on MPR. I and another campaigner responded, and although we stopped the crew continuing, we discovered that they had already amputated the branches of two other trees on the road. I was devastated and emotionally distraught; it was a horrific sight and I could not return to bed, I was so angry, and it was this anger that fuelled my next action. At 10:45 that morning, I responded to a call-out on a WhatsApp group, to say that a crew was setting up at Crescent Road to fell a beautiful London plane. There was no hesitation in what I had to do, so donning a bunny outfit, choosing clothes that I would not wear at any other time, these being a bobble hat, a 'Tree Hugga' face cover, an old pair of walking shoes and a brown mac, I made my way to the area, parking up some distance away, and walked towards Crescent Road.

Turning onto the road and looking up, I was faced with a sea of hi-vis jackets, and a large Heras fenced-off area. I was aware that my heart rate had increased significantly, my peristalsis was working overtime, but I think I was in a hyper state remembering what had occurred on MPR previously that morning, which overrode any tiredness I had been feeling through lack of sleep. That state propelled me forward. There were a significant number of campaigners cheering as I got closer, which was double-edged, as although they were supporting me, I became anxious that I would fail and not get in; the onus became even greater to get over those fences. In my mind's eye, I remember walking up and down the fencing like a caged animal, but the ironic thing was I was trying to get into the cage! I was obstructed at every turn: I considered climbing onto the bonnet of a large truck that was parked up at the corner of the fencing, and just trying to alight from there over the fencing, but a large SIA operative stood blocking my way. I knew that it was futile. I was close to tears of frustration and helplessness, but a fellow campaigner took me to one side and informed me that I could go around the back of the adjacent house that had given permission to stand in their front garden.

So, I found myself standing on their garden wall, hands on the fence looking down at SIA operatives on the other side looking up at me. The height of the top of the fence reached my hips, so it was still a challenge and I was fighting a sense of panic that I would not get over it. As I tried to raise my leg over the fence several times, with fellow campaigners who were trying to assist, I became weary and needed to take a few seconds of time out to gather myself whilst looking for a weak link. I saw an opportunity and moved quickly to my left whilst throwing my left leg over the fence, clinging to it with all my might, whilst the SIA attempted to push me back.

I found myself lying on top of the fence struggling to bring over my right leg, but I was pushed back and found myself virtually hanging off in a horizontal position on the wrong side of the fence looking into the chests of the SIA guards. I don't know how, but I think it was a fight-not-flight instinct that took over, and I found the strength to pull my body up over the fence. At this point I think even they thought that it was a pointless exercise trying to hold me and my feet were allowed to touch terra firma. Once over, my body reacted and I started to shake uncontrollably. I had developed a pounding headache and I found myself short of breath. I thought that I was going to have a panic attack and I could not move. I knew that people were talking to me but everything was muffled, I could just hear and feel my heart pounding against my chest. I stayed facing the fence, focusing on my breathing and not responding to anyone until I composed myself. The tree still stands today.

On 22nd January I was at work when I saw what was unfolding on MPR. I turned to my manager and said, 'I've got time off in lieu owing and I would really like to take it now, please'. Thankfully this was accommodated. I changed in my work car park and drove away. When I got there Amey had tried to cut off access to the work zone by putting a plastic barrier upright across the corner on the Heras barriers. I looked at it and thought, 'Aha, that's my ladder!', and in I went. I got in, and remained there for about fifteen minutes, got my five warnings, then I was manhandled out. I went limp, like a rag doll. I was unceremoniously deposited outside the fences but was back inside within five minutes. This time I stayed in longer, clinging to the fence. There was excessive manhandling to try to get me off the fence. They succeeded: I just couldn't cope with the pain, because they were bending my fingers back and causing a great deal of pain to my arm. I felt that it would snap with pressure from the SIA guy who was alongside me; he had his shoulder under my armpit, forcing my arm up against his arm which was wrapped around the top part of my arm, whilst at the same time pushing down. This was really painful, so much so I thought I would faint, but I guess the adrenalin kept me alert. I was ejected again, but I managed to get back in twice more, once getting underneath the barriers. Four times in total I managed to get inside the work zone.

On the last occasion there was a lot of support from campaigners. I was stressed by things that were being said by the SIAs, and that was making me more determined not to be ejected as I couldn't bank on getting back in again. The SIAs were saying, 'We know who you are, we've seen you before, we've got video of you'. It was a surreal situation. I tried to disconnect myself from what was happening. They tried to set up another set of barriers between me and the tree, so I was hemmed in. Another bunny came in and I was right next to them when an SIA punched them with an uppercut. I ran across the zone to two police officers and was screaming at them, 'Don't just stand there with your arms folded – please go and help, you're not seeing what I'm seeing!'. Once they ejected the other bunny, I ran back to the fence, but I think that I was very tired at that point and was forcibly pulled off and dumped on the pavement. I then became aware of a commotion and saw that the fences had come down and people were pouring in. I started crying out of relief, complete ecstasy to see these wonderful, wonderful people that had been there all throughout the weeks and months. It was just the most extraordinary moment. When I got back home I examined my bruises and had several alcoholic beverages. I was still on a high and it took quite a while to come down from the excitement and nervousness of the day. I still shake and get emotional thinking about what happened.

If I had been arrested or charged under the injunction, then I guess I would have had no choice but to deal with it. Potentially I could have lost my job, my home and my liberty. My main regret is not being available for more of the call-outs. The nature of my work means that I usually can't just leave. I regret not being able to save the trees lost on Meersbrook Park Road and on Rivelin Valley Road. Although I successfully bunnied, it's the times I couldn't do anything that I regret. I've told some friends and family members. People close to me know anyway. At work they know I'm in the campaign and they have asked, 'Have you been a bunny?'. I've said 'What, me'? So, no, I don't see the point of anyone else knowing. It just happened, it's just something that happened. There are people in the campaign who do know it was me and I've had nice things said, which is great, but there are still quite a lot who don't know. That happened naturally without me doing anything to attract comment. I don't really have a need or desire for everyone to know. I've heard it said that the injunction helped the campaign in a weird way, that it forced us to be more inventive and proactive. I'm not sure I agree: the injunction reduced the number of people doing the action of getting in the barriers. It made it harder and more dependent on a small number of people. I have a deep respect for my fellow campaigners, ordinary individuals who did the most extraordinary things!

'BUSY BEE'

I saw on Facebook the accounts of the dawn raid of the two retired ladies on Rustlings Road and I wanted to get involved. I'd heard about a meeting at the Hartley Street Institute so I went along, but left after twenty minutes because it was just a backbiting party-politics argument between Labour and the Lib Dems as to who was more involved or to blame. Very boring.

One day sometime later I saw a lady on the streets in Norton with a clipboard so I approached her. It was Celia Pinnington and she put me on the mobile-phone call-out group. The first call out that I was able to respond to was to a huge, majestic sycamore at one end of Lees Hall Avenue. There was fencing up around the tree, but there were two ladies under the tree who had gone through someone's front garden to get there. So I went through the garden too. I stood with a lovely lady who I had only just met; it was her first time protesting too. The resident, a young woman, came back home to her house accompanied by an old man, a neighbour. She came out and asked us to leave. I apologised and said I was not going to leave. I explained that I was not on her property, we were on the pavement, and so we were not trespassing. She said but you did trespass to get there; I admitted that we had and I apologised for that. She went back inside and phoned the police. I think she was being coerced by the older neighbour. A very nice woman police officer came and asked us to leave. I asked if she was going to arrest us and she said no. I said we were not going to leave. We stood there for about four hours. One of the Amey guys, Jay I think, spent most of the time talking to us, telling us his life story, about his ex-wife, etc. Then they took the fence down and we all left. In the night somebody 'ring barked' that sycamore and killed it. So it was felled a couple of weeks later.

My next big adventure was at the other end of Lees Hall Avenue. I approached the tree with the fencing in place around it, and by observing what happened previously, I had realised by then that you had to move very fast, to get in place as quickly as you could. It was not effective standing outside the barrier; you had to be as close to the tree as possible. By the time I got there, the barriers were incomplete, so I was able to quickly jump over the smaller fencing to get in place. I was outside the fence, but not in the garden, just between. As soon as they realised what I was doing, an Amey guy, a big bloke, came over and stood right next to me, pressing the fence into me. The lady living in the house adjacent to the tree came out and asked me to move. I said I was very sorry but I was not going to move. She said her husband was in the house and he was terminally ill and dying, and I was upsetting him. I said I was very sorry if that was the case, and I would go away when all the arbs and Amey went away. She said that the residents should be consulted and that they wanted the tree removed. I said that her neighbour in the next house, who was standing right there near me, did not want the tree removed.

The neighbour in the house on the other side came out and started verbally abusing me, shouting and swearing and threatening to turn the hose pipe on me. It was very cold so I was glad that he didn't. The police arrived and asked me to move, saying again that I was upsetting the dying man in the house. I apologised again, saying that we would all go away if the arbs and Amey went away first. I politely pointed out to the police officers that the only thing that was protecting that beautiful ash tree from being destroyed was me standing underneath it, and so I could not just go away and leave it. The police left and a while later everyone packed up. I went to bed really anxious, hoping that the old dying husband of the shouty lady didn't die in the night. I would have felt awful if that had happened, but it didn't. That ash tree is still standing.

Many of the actions I did were in the cold during early 2018. On Meersbrook

Park Road one such day the crews were setting up the fencing around about five trees. Suddenly another bunny leapt over the Heras fencing to protect one of the lime trees by the park entrance. I walked down the road towards the other end. There were two very respectable and polite gentleman protesters squashed in between some orange plastic fencing and the wall by the tree. The SIA security guards were aggressively pushing the fencing against the legs of these two gentlemen who were trying to push it back to relieve the pressure on their legs. I thought, this is not right, there is an injustice happening here. So I quickly jumped in my car, drove home, changed into all black clothes, put a hat on and a scarf over my face. I dashed back, parked round the corner, and just walked really quickly and calmly towards the tree where the gentlemen were still being assaulted by the security guys. Before anyone noticed I had stepped over the orange plastic barrier and sat under the tree. My heart was pounding, my mouth was dry, I stared at the ground and when they came to ask if I was going to move I shook my head, careful not to actually speak. We stayed like that for a couple of hours, a stalemate. Then they all started to pack up, so I escaped from behind the barrier and wandered down the road to find the other bunny. He said we had to be really careful as they try to follow you. We had just taken off our masks and coats and I was going to double back to get my car. Just as I came out of the alley there was a security guard there, who had followed us. I don't know if he realised that I was the person in the disguise or not, but he went down the alley to follow the other bunny, but did not follow me. I phoned him to say he was being followed, so he picked up his pace and made it into some nearby woods. At this point the security guy gave up the chase.

Another day on Meersbrook Park Road it was so cold, and stuck under a tree for hours it was even colder. You get offered hot drinks by sympathetic neighbours but if you drink then you need the loo, so it is best not to drink if possible. It is also best to keep changing what you wear so that you don't get recognised. I decided to arrive at the site in my disguise. A bunny was already over the fence and ensconced under the tree at the entrance to the park. I was wearing a bright orange hi-vis coat as it was the biggest coat I had and could fit several jumpers underneath. This was a mistake as it was the same coat I had worn on Coverdale Road when I did not have a mask on. One of the security guards, a short Scottish bloke, started calling after me, 'I know who you are, I recognise you, you drive around in that little car'. So I realised that I could not do anything risky that day, like getting inside the barriers, but I kept my mask on anyway. Several memorable things happened that day. There was a big crowd of protesters, many of them retired ladies. I noticed that in fact the fence was not secured, and there was no cable tie holding two sections together. When I thought nobody was looking I tried to pick one section up to get inside, just to cause a distraction. Of course I did not get very far as there were so many security guards, who pushed back hard, knocking several of the lady protesters to the ground. Later on when they tried to make an official complaint about this, they were told, 'Well what do you expect? If you are at a protest you cannot be surprised if you get injured'. Meanwhile at the other end of the street there were two bunnies under one tree. The police had been called and they were on their way. At intervals the security would go to ask them both if they would like to leave, and after a while one decided he did want to leave and was escorted out. The police arrived and apparently they had been told that one of the masked protesters had bent the fence when he had jumped over it, committed criminal damage and was going to be arrested. It turned out that it had been the bunny that had left. They tried to persuade the other bunny to leave but he said no. The crowds outside the fences were all jostling and shouting. The police got hold of the remaining bunny and started dragging him off. This was the first incident that I had seen where they didn't just try to negotiate with a bunny but actually physically dragged him away. They dragged him outside the fence and dumped him on the ground, but in the melée and jostling he somehow managed to get under the Heras fencing and back inside the work zone. I guess the police were fed up with him by then so they dragged him out again and over to the police van. A police officer carried each arm and leg but faced him downwards which looked very painful. After he had been taken away in the police van I was still on the street with my big orange coat and mask on. I suddenly noticed that my fourteen-year-old daughter was there watching as she was walking home from school. I did not approach her as I thought she'd be shocked or embarrassed. Later she told me she had been there and the security guards had told her that she shouldn't be there, that she shouldn't be watching, which she ignored. I told her that I was there and she shrieked, 'Oh my God, you were the one in the orange coat!'. You have to be very careful with your disguises.

One of the amazing things about the whole protest was that it was so spontaneous. There were some protesters more vocal and more active than others, but there were no actual leaders. People just did what they felt was necessary and what needed doing to save the trees, but as the months went on they got braver and more actively involved, standing up for what they believed was right. Another day on Meersbrook Park Road there was fencing around several trees – massive enclosures to keep protesters out as well as machinery, arbs, SIAs, a huge operation. There were five protesters, bunnies who had hopped over the fence, one under each tree. A stalemate, which means victory for us. Five people were willing to go through the discomfort of masking up, being terrified, breaking the injunction, going to court, going to prison, to do what was right and save the trees.

22nd January 2018, the battle for Meersbrook Park Road, was my birthday. I was at work, but I always had the live-streaming films on my phone and realised that it was all kicking off. I did not have my mask, but I only worked ten minutes away so I just left the office. The whole road was cordoned off. Heras fencing erected around the trees all the way along the parkside edge. I quickly approached and saw a retired lady and her husband, both in their eighties, walking with sticks, standing under their favourite tree opposite their house. There was a terrible commotion going on, which I later found out was because a security guard, known afterwards as 'Uppercut' had punched a protester and everyone was focused on that. The fencing near the retired couple was open so I just skipped up and linked arms with them. The security guys didn't know what to do as they could not touch the retired couple or me. I realised that I should move to be under a different tree as this one was well protected, and I was already within the fencing so I could do that. I was not breaking the injunction because the fencing was incomplete. I ran along and linked my arms through some Heras fencing and under the next tree. Security came over, there were about eight guys in hi-vis surrounding me, with one younger lad trying to quite viciously prise my fingers off the fence. I was shouting, 'Leave me alone. I am a respectable, middle-aged woman and it is my birthday today!', which they thought was quite funny.

We were there for some time, all jostling, pushing, shoving, shouting, until someone realised that the section of fencing I was holding on-to was not an integral part of the barrier, it was an extra piece, so they cut the cable ties [and] I was pushed out of the enclosure. Meanwhile there were three people, three young women I think, all masked up, who were taking it in turns to jump over the fence from the park into the enclosure. They would sit under a tree, security would caution them, ask them to leave and then drag them off, literally drag them through the mud. As each one was dragged off, another would jump over the fence. It was brilliant. I was inside the park by the fence at this point, whilst one of the women was by the fence, holding my hand on the other side to keep from being dragged off, but the security were very aggressive and determined to drag her away. I was shouting at them, 'Don't hurt her! Don't hurt her!'. All the protesters outside the fence were shouting and pushing the fence and suddenly it collapsed. Just as this was happening I noticed that the loose section of fencing that I had been holding was just in front of me; at the same time as the perimeter fence collapsed I gently nudged this fence and it crashed to the ground too, adding to the general noise and confusion. Then everyone ran inside the enclosure to rescue the women protesters and to link arms under the tree. I ran round to join them. As we stood there a security guard said to them, pointing at me, 'It's her birthday, you know'; so, linked arm in arm under the tree the protesters sang 'Happy Birthday' to me. It was a great moment. Then, to add insult to injury, because they had been so emboldened by their success, the protesters stood in front of the Amey vehicles as they tried to leave, making it a very long and fruitless day for them.

By that stage in the protests there would be ten arbs, twenty security and evidence gatherers accompanied by up to thirty policemen every time they tried to fell a tree, which ramped the whole thing up to an unbelievable level. It was a civil war waged on our streets.

One other incident that makes me laugh to think of is when my daughter knew I was going to a tree-felling. She texted me saying 'We have tickets to go to see *Mama Mia* at the Lyceum tonight and I have been looking forward to this for months. DON'T GET ARRESTED!'.

'JUMPING JEN'

After the first injunction, the change of mood was shocking. No more chatting, no more shared tea. This felt more like war. The horrors of Heras fencing and masses of security guards and police were still to come, but the rules of engagement had changed and the mood was bleak. I remember one morning, it was the week when Sheffield was essentially cluster-bombed with arbs, many we had never seen before. The panic on Facebook and WhatsApp was terrible. People calling for help, not enough people to cover all the sites. We were losing and drowning. I was going to work. Work was no relief as sitting at your desk, with the pleas for help rolling in and not being able to help was even worse than being out on the streets and at least doing something. So I ran to work. It was icy and I had a nasty suspicion that they would try to hit trees in the valley bottoms and near main roads because of the ice, so I ran out of my way via the entrance of the Botanical Gardens. A hunch, and there they were, just setting up at Thompson Road, fencing off one of the lovely trees at the entrance to the Botanical Gardens. Small plastic fences around three sides of it (we now know this wouldn't have classed as a complete work zone, but I didn't know this at the time) and a huge sign instructing me about the injunction. Chippers, work vehicles and very unfriendly-looking arbs. And there was no one else there. I had bought a squirrel mask from eBay a few days before and still had it in my running bag, ready. I was relieved to find I hadn't unpacked properly. It was still there! I didn't think any more, I just couldn't see that lovely tree felled. So I pulled on the mask and ran up, got in the barriers and sat down. I sat mainly because it appears that latex squirrel masks have very limited visibility and I couldn't trust my feet on cobbles, and I had already walked into something hard. It's not a nice thing, arbs shouting abuse at you and taking 'evidence' photos of you. It's cold. For some reason it always seemed to be winter in the tree campaign. My Raynaud's Disease means my hands go ridiculously numb really quickly, and this day was no exception. It's cold, it's lonely, and you don't have a clue what's happening as staying logged into your Facebook and WhatsApp feed seems somewhat foolish when you are trying to be anonymous in order to avoid court action. A quick update on WhatsApp as to where I was and what was happening was all I could manage, and even that felt risky. On this particular day there were so many crews throughout Sheffield, so our support was stretched.

Russell [Johnson] appeared for a while and I was glad to see him, but otherwise I was on my own. My hopes that I might get the word out and be able to get off to work just a bit late were gone. So I just had to sit. It's timeless sitting under a tree. Minutes slide by so slowly, but the hours are suddenly gone. It's boring and terrifying. Eventually the arbs go. It's an agonisingly slow process taking down the whole tree-cutting circus, and you can't leave until they have all gone. Otherwise they would just start up again. I remember running back out through the Botanical Gardens worrying about being followed, as by this point rumours of people being followed home were about (and as it happens, true). I took off the mask as soon as I was out of sight and off the main path. Luckily I had a spare running top in my bag, so I was able to exit the park as a differently clothed runner. I still did a long and merry run with many switchbacks before eventually arriving very, very late to work. Having a boss who was a firm supporter of trees certainly helped. It's an odd feeling, thinking you might be being followed; a bit frightening, but also slightly thrilling. You are on your home turf, and you are a long-distance runner. No one will keep up with you on foot, and you can go places a car cannot. You just need to plot your route. That tree is still there. I've spent more time and taken more risks for ones that have gone, but that one is there. Other people have saved that tree too, but if I hadn't done what I did on that day, it would be gone and when it's gone, that's it.

'GRANNY BUNNY'

Since that awful night on Rustlings Road I have been to many tree protests to observe and to take any peaceful actions that were open to me. I have stood for hours with others in the freezing cold, refusing to move when asked. I have been on marches, raised funds to help pay for the campaign, become very emotional when seeing a tree being cut down after trying to save it, and made many friends along the way. The campaign to save Sheffield trees has brought out of the woodwork so many amazing and determined people; people who have been prepared to put many hours into the campaign, who have written letters, given speeches, raised funds, used their expertise to help in whatever way they can. But I have also seen many ordinary citizens prepared to just stand there and not move because they knew that moving away would mean yet another tree being cut down.

I realised one day, while standing outside the low plastic barriers round a tree that was about to be felled, that the only way I could immediately stop the felling would be to climb over and stand under the tree. Once I had that thought, I couldn't let go of it. Really? I wanted to be a 'bunny'? I had seen others do it and knew it worked, as at that stage the arbs were not allowed to touch anyone (that changed, of course, especially once the campaign escalated). I didn't do it straightaway, but went home, thought about it, decided I was mad but would do it anyway. I then had to choose my wardrobe – mostly what to wear to cover my head and face as much as possible. The first time I bunnied my heart was pounding loudly when I climbed over the barrier, but then everything immediately went quiet. The arbs downed their tools and I just stood there doing nothing. And stood and stood, because I couldn't leave until they had packed up and moved on – which they did. I knew I had to do it again, which I did, several times. But then the barriers got much higher and the fight against the tree protesters got much uglier. After a few more bunnying experiences I had to give up when I saw other bunnies being forcibly ejected from the work zone. I was afraid of being injured and besides, it was impossible for me to scale the tall metal fencing that was now being erected around the trees. I regularly saw other bunnies in action, but never knew who they were, and it was quite a surprise to discover some of their identities later on. Even now I don't know who they all were, as we all needed to protect our identities, for obvious reasons. Anyway, I don't care who knows now. I did it for the trees and would do it again.

The campaign to save the trees has always been about more than just that. It's also about standing up for what is right and not allowing ourselves to be bullied and walked over by people who should have our interests at heart but who refused, over and over again, to listen to us. The campaign is about democracy and transparency and accountability. The campaign is about everyone who cares standing up and refusing to sit down again.

'ROB-IN-HOOD'

I didn't worry about what was going to happen when I sat there under the truck. By this point in the campaign I was aware the police would be taking my details and passing them to the Council. But I was fed up, I think we all were. The weather wasn't as warm as it had been in the summer. The barriers set up around the fellings had grown taller too. Gone were the traffic cones strung together that you could step over and cease an unnecessary felling with ease, replaced by Heras fencing and hired thugs in yellow jackets, ready to put the boot in if you should try and stop them erecting a safety barrier. For a while it seemed they got their way: that we were only delaying the fellings, safeguarding the odd tree on a road, whilst the rest were taken. It was draining work, slogging round patrol routes trying to find the sites before they had the fences up, or racing to an alert from the 'Flying Squad'. I can't even imagine how many hours some put in. I hadn't, but I'd been around enough to be in some scrapes. My thumbs had been twisted off the railings at Meersbrook Park Road, and I'd been shoved over and kicked at the entrance to Thornsett Road in the snow. Why had I done this? Hearing the stories of several generations of a friend's family playing under the chestnut on Chippinghouse Road, and seeing them cry as the majestic tree was felled and chipped in front of their eyes. I felt the urge to run under its great boughs then, but the sobering words of Chris [Rust] tamed that impulse, 'We can afford to cover the fine, so far. Might be worth waiting...'. He muttered something about making an impact. It took months till that moment came. I'd been led up the garden path on the morning in question. A wagon of security sat idling on Abbeydale Park Rise and I'd fallen for their diversion. Chris Burn from the *Yorkshire Post* had interviewed me, anonymously as a masked bunny, before a message came through.

They'd set up on Kenwood Road, and I'd left my transport (one of those hire bikes) back down by Millhouses Park. Walking back along Abbeydale Road towards Nether Edge a car pulled up and I recognised Justin [Buxton] at the wheel. 'Fancy a ride?', he grinned. Arriving at the scene, there was already a crowd of protesters surrounding a large section of Heras fencing. There was an almost carnival-like atmosphere, as the pantomime of police, protesters, arbs and thugs, all played their respective roles. I found myself observing the situation from a stone wall I had climbed up to erect a banner from.

The felling seemed imminent as two geckos were walked out of the safety zone. I looked up and made eye contact with a fella I had met by Meersbrook Park in the summer. I'd been walking home when the sounds of some psychedelic rock wafted down from the canopy of a tree that day and I noticed a protester sat high up in its branches. "Ere, you wouldn't fetch the ladder? It's stashed at number 57', he'd called down to me one summer's day. Now, here we were on a relatively mild day in March. A slight nod was enough; I knew at that moment we were jumping over. I didn't know what we planned to do, and neither did he. All we wanted was to delay them in the hope we could come up with something, a plan, or a mass movement, to save the special trees on our roads. The rare pink-fleshed apple tree I knew was on the list, or the great old oaks sat along ancient borders, now due to be chopped because they intervened into a road by an inch or so. We didn't get very far before the security mobbed us. I was expecting to be roughed up a bit like on other occasions, but they held back. The thugs were nervous, probably because a police CCTV truck was parked adjacent to the scene pointing a camera right at us. 'Don't let him get under the truck', I heard one hi-vis wearer say to another, so that was immediately what I did. Crawling under the cherry picker, I was suddenly free from their grasp and at the same time trapped: confined to the underbelly of this horrible machine, bought here to allow the arbs to reach the upper branches of the tree with ease. I remember the stink of diesel fumes and the cold tarmac slowly numbing my mind. Although there was quite a crowd of protesters only feet away, at that moment I felt isolated. I heard shouts after an hour and a half as my partner in crime was carried out and nicked. I remember having the forethought to go through my phone and delete anything I thought might be compromising. As I was doing this I heard shouts from the crowd, 'Over your shoulder!'. I turned to see a hired goon trying to film the screen of my phone from behind. At the time I was deleting pictures of friends from a rather boozy holiday in Benidorm, hardly the smoking gun they were hoping for. The police came and asked me to leave peacefully, but said I would be arrested either way. I guessed sitting there would buy more time and I would be £200 down either way, so I stayed sat still, playing the waiting game. I heard Simon [Crump] climb in a nearby bush at one point, shouting a mixture of encouragement and abuse, and goons came back with cameras, this time offering a foil blanket.

Their attempt at compassion rang about as true as the Council's claim there was no target to cut down trees, despite many being on the list for spurious reasons.

After nearly three hours, things seemed to be drawing to a finale. The riot squad had arrived and were suiting up. Jacks had been placed under the axles of the truck, and the nearby protest had swelled. A face in the crowd called for me to pass them my phone and I slid it quickly along the ground before the riot squad formed up. I tried to stand my ground, but was quickly overwhelmed, dragged out and de-masked before being thrust towards the cameras of the CCTV wagon. The coppers seemed disappointed it was me, like they expected it to be someone else. The line fed from the establishment to the media was that there was 'a small number of hardcore protesters', so I imagine they expected me to be a familiar face. As I was carried out, a fellow protester thrust a card into my, by then, cuffed hand for the solicitors: 'Use these mate', he had time to say before being shoved back by the Filth. As they closed the doors of the paddy wagon, I could hear the protesters outside, some singing, some cheering, others jeering; a clearly identifiable accent was shouting about Orgreave. I felt defeated despite the audible support. Roughed up, shaken and overcome, I began to cry. What was the point? They'd beat us. They were too powerful, we couldn't stop their machines from tearing down our trees by the thousand. I spent the evening in the cells waiting for some detectives who outwitted me in the interview. 'No comment', the solicitor firmly mouthed as their conversational tone nearly tripped me into talking after having been alone in a cell for hours on end. It was probably a well-worn tactic of theirs. Luckily it all ended that same day and I was released in the evening with the solicitor advising me to maintain a low profile and not to fall for getting into a chat with the officers when they offered me a lift home. I spent a night alone and with regret. The arrest could cost me my career, a conviction could set me thousands of pounds in debt. My coat was ripped and the next day at work questions were asked as images of me being carried off by the riot squad were on the telly. This whole thing seemed to be picking up a lot more momentum than I had expected. That evening, Bridgette, the woman I had wept with while her favourite chestnut was taken down, arrived at my flat with my recovered phone, a cake and some beers. She wished me well and the significance of how many others had stood behind me in that action really sunk in.

It was a real community resistance. Generations of people had loved the many trees around Sheffield each in their own way for many reasons. From the tree they grew up with, to the home for local birds, butterflies, bats and insects, to their shade on a summer's day walk, to the wonderful green canopy they get to experience every year from early spring, through to the autumnal crown of golden brown. Two days later, my image was repeated on the BBC news cycle all day. An exposé by the *Yorkshire Post* had blown apart the claim of there being no set target of tree replacement in the contract, and suddenly the fate of Sheffield's street trees was not just a matter for local folk but had taken on some sort of national significance. A resistance that only days ago had left me feeling so personally defeated now took on a new hope.

AFTERWORD

On 21st March 2018 a team of arbs, barrier crews, security goons, and police, turned up on Rivelin Valley Road for the second day running. South Yorkshire Police were in fine form, putting a man in his mid-sixties in a chokehold and arresting a woman for blowing a plastic trumpet. The next day, they followed it up by arresting two more women for fearlessly brandishing a tambourine and a pink sparkly recorder. But the third in a trilogy is often the most disappointing, and by Friday all they could manage was to facilitate the felling of two Second World War memorial cherries and the random arrest of a man who was dancing to justify their presence that day. And then it stopped, for Easter and the Council elections, the 'dangerous' tactics of the protesters on Meersbrook Park Road cited as the reason.

We waited for it all to start again after Easter, but it didn't, and as the weeks turned into months we started to hope that this was really IT, that we'd actually stopped it. We didn't know at the time that the Forestry Commission had started to investigate our allegation that Sheffield City Council were felling hundreds of trees without a licence, a legal offence which carries a significant fine. To keep ourselves amused, we started a book on how many times the then Cabinet Member for Environment, Streetscene and Climate Change could use the phrase 'the way forward' in a conversation, and when SCC were going to unveil the proposal they said they'd present to us. We asked to be involved in drafting it, but unsurprisingly this was refused.

As spring turned into summer, the late Roy Millington, an old soldier now in his eighties, began his own protest, sitting on the Town Hall steps in defence of the First World War memorial trees on Western Road, Crookes. We didn't know at this point that Amey's proposal included saving these trees, and I wonder how many times SCC officers and councillors walked past Roy's dignified presence, knowing that they could have said, 'Go home; you've won'. Eventually, someone at SCC cracked, and in early November leaked the information that all but three Western Road trees were saved in the 'proposal'; later the remaining three were saved after investigation.

In late September we sat down to an initial three days of talks with Amey, Sheffield City Council, a mediator and the chair, Bishop Peter Wilcox. We were told that we'd present our cases to one another before we started. We sent ours as requested and Sheffield City Council didn't, so we went in cold and reacted to what we were presented with as we'd done before on the street. We've never been released from the agreement to keep what went on confidential, but I can say that we went in asking for resolutions for the past, the present and the future, in the form of an inquiry, the current felling to stop and be replaced by proper appraisal, and a street tree strategy to be produced so street trees were valued and cared for in future. It took some tussling, but we came to a mutual conclusion on all but the inquiry, SCC asserting that they were 'not minded to do this at this time'.

On 15th January 2019, a reasonably large crowd gathered on Chatsworth Road for the first of the 'Joint Inspections', a process by which a specialist Amey crew would attempt to repair the 'impossible' trees on the day, in the presence of STAG observers. Amey were very nervous, though to be fair there were so many of us milling around that they were right to be afraid that one of us might get run over. The tree was saved in a few hours, which was pleasing and intensely irritating at the same time, since it was clear that the others could also have been saved if the will had been there to do it. This was the first outing of the 'Pink Panthers', named after our natty hi-vis outfits. We'd been told that if we were going to be near/in the safety zones we needed to be able to be seen from space and drew the line at donning Amey gear, so we ordered our own and added a new animal to the STAG menagerie.

More trees continued to be saved by an increasingly enthusiastic and innovative Amey team, though some roads have needed much more work and a complete redesign in order to keep the existing trees. Also, there are those trees that had their own day in court, courtesy of those that defended them; trees that are now saved but were the cause of court cases for breaches of the injunction gained by asserting that the highway could only be properly maintained by removing them.

Almost exactly a year after the talks, in September 2019, the Street Tree Strategy Development Group met and started to draft the street tree strategy, and after a lot of hard work and some very frank discussions this was approved in final form in March 2021. It contains the intention to increase, improve and maintain our street trees, so a return to the past isn't possible within it. It's genuinely a tremendous piece of work which all parties should be proud of, and testament to what can be achieved when people are prepared to leave their baggage at the door and meet as equals to do the best they can for the city.

So, that's the present and the future covered, but what of the past?

Sadly, at the time of writing, and despite two damaging Ombudsman's reports about street-tree policy, Sheffield Labour Councillors are still unable to admit that they might have got this so badly wrong. We've never had an explanation as to why they thought getting an injunction and trying to put their own citizens in prison was the right way to go. We've not had a clear answer on whose decision it was. There's an increasing amount of denial over these last points but 'In the application the Council will be seeking your committal to prison' is hard to write off as a typo.

It was expensive too; we'll never know how much was spent, but over £400k of legal expenses has been identified, only some of which has been recouped through legal costs against those found in court to have breached the injunction, and SCC had to pay compensation of £700k to Amey for delaying the tree programme with the Independent Tree Panel process that they mostly ignored when it said 'don't fell'.

On top of those countless staff hours were spent on this whole debacle that could have been put to better use. Let's face it, if they'd given the entire Council three weeks' holiday in the Bahamas it would have been better value.

Is there anything else? Yes! We made street trees newsworthy and encouraged other protest groups. Some councils gave up without a fight, saying, 'We don't want to do a Sheffield, and started to talk to their residents. We even changed the law of the land; the draft Environment Bill includes a 'duty to consult' on street tree-felling, which has been dubbed the 'Sheffield clause'. That's something to be proud of, if it makes the final cut.

Was it a complete success? A good question. There are still many people haunted by memories, every time they step out their front door, of a tree which is no longer there. It has taken years for some of us not to be triggered by the sound of a chainsaw or the sight of hi-vis. Three years on, we are still healing.

It was emotionally and physically hard, we fought hard, so did they, and we lost thousands of trees.

But we saved thousands more.

Christine King
Co-Chair, Sheffield Tree Action Groups (STAG)

PRINCIPAL PERSONS: AKA THE OPPOSITION

Over the course of the protests several people from Sheffield City Council, Amey, the Labour Party, South Yorkshire Police or subcontractors, became well known to protesters. Usually this was done by being a public figure, or becoming one by actions such as giving evidence in court against tree protesters.

Barrett, Dom – An Acorn arb who became well known to protesters.

Billington, Paul – Sheffield City Council Officer placed in charge of the tree-felling policy in 2017. Gave evidence in both injunction trials as well as the committal hearings in which Sheffield City Council attempted to imprison protesters. Took early retirement following the ending of the felling programme.

'Bob' – One of the SIA operatives who was often the target of Benoit's poetry.

'Boris' – An evidence gatherer. Named after bearing a passing resemblance to Boris Becker.

Buck, Andy – Chair of the Independent Tree Panel, set up by Sheffield City Council to advise on trees scheduled for felling.

Butt, Darren – Amey Accounts Manager, who gave evidence in both injunction trials as well as the committal hearings. Regularly appeared in the media defending the policy. Also regularly seen at protests.

Dore, Julie – Leader of Sheffield City Council throughout the protests until her retirement in January 2021.

Ellison, Carl – Amey arb supervisor regularly seen at fellings. Gave evidence in court cases.

Fox, Terry – Sheffield City Councillor. Cabinet member responsible for the Streets Ahead contract between 2015 and 2016. Became Deputy Leader of SCC in 2019 and Council Leader in May 2021.

Haigh, Louise – MP for Sheffield Heeley. Has a constituency office round the corner from Meersbrook Park Road.

Lodge, Bryan – Sheffield City Councillor. Cabinet member responsible for the Streets Ahead contract between 2016 and 2018.

Males, Stephen – High Court judge who presided over the first injunction hearing in 2017 and the subsequent committal hearings.

Robinson, Graham – High Court judge who presided over the second injunction trial in 2018.

Stubbs, Ian – South Yorkshire Police inspector who oversaw the use of Trade Union law against protesters. Retired.

'Uppercut' – An SIA operative notorious for punching a protester on Meersbrook Park Road.

Wignell, Jason – An Acorn arb often at the centre of clashes with protesters. Gave evidence against several protesters in court.

GLOSSARY

Acorn Environmental Management Group – Tree surgery company which carried out many contentious fellings whilst subcontracted to Amey.

Anderson Tree Care – Locally-based tree surgery company which carried out early work in the contract whilst subcontracted to Amey.

Amey – Company contracted by Sheffield City Council to fell ('replace') 17,500 street trees over the course of the Streets Ahead programme.

Arbs – Arboriculturalists.

Barrier Crew – Workers responsible for the erection of barriers around the trees before felling.

Bunny – A protester (usually disguised) who jumped into work zones to peacefully prevent a tree-felling following the High Court injunction.

Bouncers – Commonly used term for the SIA guards who staffed the protests after the injunction was awarded (see SIA).

Cavat Value – Capital Asset Value for amenity trees. A system used to give an approximate asset value to a mature tree.

Cherry picker – (see MEWP).

Chipper – A distinctively shaped machine which converts branches into woodchip, the sighting of which sparked many an alert.

Dronfield Tree Services – Tree surgery company subcontracted to Amey early on in the programme. Widely criticised for their practices.

Evidence Gatherers – Staff employed to photograph and identify protesters both in the build-up to the injunction trial in July 2017 and subsequently.

FOI – Freedom of Information request.

Garden Permissions – A tactic used to great effect by protesters. Standing in a garden or driveway with permission of the resident to get close enough to a tree to prevent its felling. First used on the final threatened Rustlings Road tree and then systematically employed across affected parts of the city.

Gecko – A protester who got close enough to a safety zone to prevent barriers being erected (or for felling to take place). Named for sticking to walls.

Gnomes – A protester who got close enough to a safety zone to stop work by standing (or sitting) on walls.

Heras – High metal barriers used after the injunction to prevent protesters accessing the safety zone, not always successfully.

Hi-vis – Collective term given to the people staffing and policing protests, named after their distinctive high-visibilty yellow jackets.

Injunction – Restraining order first made against protesters in August 2017 to try and prevent people entering the safety zone, not always successfully.

MEWP – Mobile Elevated Work Platform. (see also Cherry Picker)

Night Raids – A tactic used later by Amey to remove branches during the night to make daylight fellings quicker and easier, not always successfully.

NVDA – Non-Violent Direct Action.

Oversail – A disputed point of law never tested in court. Protesters believe that residents could deny legal access to people working over their property.

PFI – Private Finance Initiative. A controversial method of privatising public services which was used for the Amey contract.

Photo arb – Before the evidence gatherers, one or more arbs were sometimes employed in the gathering of evidence.

Process Server – Employed by Sheffield City Council to deliver and serve legal documents on defendants.

SCC – Sheffield City Council.

Shepcote Lane – South Yorkshire Police local detention suite where many a peaceful protester was detained.

Streets Ahead – The name given to the twenty-five-year highway programme starting in 2012 which Sheffield City Council contracted to Amey.

Six 'D's – Dangerous, Dead, Diseased, Dying, Damaging, Discriminatory. The categories that condemned trees were put into by Sheffield City Council.

SIA – Security Industry Authority. Accreditation body for security guards.

SNET – Save Nether Edge Trees

SORT – Save Our Roadside Trees, formerly Save Our Rustlings Trees. Early campaign group based around Rustlings Road, which successfully raised the profile of the issue in the early days of the campaign.

STAG – Sheffield Tree Action Groups. The umbrella group of local tree campaign groups in the city.

Squirrels – Protesters who prevented fellings by climbing trees.

SYP – South Yorkshire Police.

TULCRA – Trade Union and Labour Relations (Consolidation) Act 1992. Controversially used by South Yorkshire Police in conjunction with Sheffield City Council to arrest tree protesters between November 2016 and February 2017. All cases collapsed before trial and were the subject of out-of-court settlements by South Yorkshire Police to those arrested.

Undertaking – An offer made by Sheffield City Council to protesters that they promise not to take part in direct action anymore and therefore avoid the legal costs of being taken to court. The refusal by all defendants to sign in 2018 was described in court by Justice Robinson as 'eminently justified'.

Vampire arbs – Arbs carrying out night raids.

THE LEGAL STUFF

PAUL POWLESLAND

'I really enjoyed having Calvin as a client, mainly because he didn't do the thing that most protesters do when they get into court: 'Oh shit, I need to get off now!'. His thing was, 'No, I did it for the cause, and I'll do whatever is useful for that end'. After that, every court case strengthened the campaign and weakened the Council. I didn't get the easy victories I had early on, after Chippinghouse Road, but I was so privileged to be a part of it. We found things out that the campaign could use in future, things said under oath that could then be asked about at Council meetings, for example. Protest slows down irreversible damage and gives time for the legal process to happen; it creates that opportunity. I've rarely known environmental campaigners who have had such clear success. Sheffield tree protesters achieved their aim; through some of the harshest legal things being thrown in your direction you stood up and you won!'

Court Reports by Sheldon Hall, tree protester

FOR THE AVOIDANCE OF DOUBT: THE CASE OF SCC vs STAG

On Wednesday 26th July 2017, twelve angry men and women took their seats in a jury box at Leeds High Court. They were not, however, members of a jury. If they had been, the outcome of the case played out before them would have been quite different.

They were instead present as supporters of the Sheffield tree campaign, and of the nine campaigners named in Sheffield City Council's application for an injunction to prevent the non-violent direct action that was successfully holding up the felling of street trees in the Streets Ahead programme. Observers were permitted to sit in the jury box to free up space in the public gallery, such was the interest in the case. This account is based on the notes taken by the present author, from the vantage point of one of those jury seats.

The case was heard by Mr Justice Males and scheduled to last three days. Before the first session had begun, six of the named defendants had signed undertakings agreeing to abide by the terms of the injunction application. John Cooper QC, defending, noted that they had signed under pressure because of the cost implications, and had felt intimidated into doing so. The counsel for the Council, David Forsdick QC, insisted that they had signed of their own free will. Justice Males ruled that they would be bound by the undertakings as if an injunction had been served and would be liable for contempt proceedings if they were breached.

The injunction sought to prevent protesters from entering a safety zone around a tree or encouraging others to do so, or from deliberately obstructing tree works. The tenth defendant in the application was named as 'Persons Unknown, being persons intending to enter or remain in safety zones erected on public highways in the city of Sheffield'. At stake, therefore, were the rights of the general public to protest *effectively* against actions that they believed to be (in the words of the Judge) 'inappropriate, unnecessary, and unjustified'. If the injunction was granted, anyone in the city would be bound by the same restrictions as those who had signed the undertakings, and faced with the same consequences if found to be breaking them.

The two counsels were a study in contrasts: David Forsdick, tall, with a floppy brown fringe and an air of quiet self-assurance; John Cooper, short and stocky, with close-cut grey hair and a more confrontational, even theatrical, manner. Off stage he spoke with a light Wolverhampton accent and apologised for not having the time to be polite. The lawyers had encountered one another in a previous case, when Forsdick had prosecuted and Cooper defended Occupy London over their 2011 protest camp on the steps of St. Paul's Cathedral; Forsdick had won.

The remaining defendants who had not signed undertakings were Alison Teal, David Dillner and Calvin Payne. None of the three denied involvement in direct action, though Justice Males remarked, 'They didn't like the phrase'. Mr Forsdick said that the Council accepted that the protests were not violent and the protesters' motives were sincere: 'A degree of tolerance by public authorities is required of lawful, peaceful protest'. But in this instance he disputed their lawfulness.

Mr Cooper opened with a central question in the case: what influence was Amey having over the Council? He argued that, 'Amey brings down trees in such a way that it is heavily induced for them to make profit, and to avoid alternatives that stand in the way of their making profit. Healthy trees are being felled to increase this profit and make good on losses caused by the past neglect of the Council. There has not been a proper assessment of the trees brought down, or about to be brought down'.

Key to the case, he argued, was the nature of the PFI contract with Amey, which was available for inspection only in heavily redacted form. Mr Cooper noted that it was, 'intriguing because no-one has seen it – even the councillors'. Opposition councillors had made repeated requests to see the contract, but they had always been refused. Was there a provision to penalise the Council if the project was delayed by protesters, as had been claimed? What were the provisions for felling? Mr Forsdick responded that the contract was a red herring. Getting value for money was not improper but an obligation on local authorities. How it went about fulfilling its obligations was up to the Council, and if it wanted to cut down all the street trees in Sheffield it would be entitled to do so.

Mr Cooper said that the issue was whether the Council was making its decisions properly and proportionately. He did not accept that the Council was merely fulfilling its statutory obligations, under Section 41 of the Highways Act, to maintain and repair the highways. The defendants demanded accountability and transparency and 'at the very least' the judge should see the contract. However, Justice Males was 'not minded' to see something that other parties in the case had not seen and 'not persuaded' that disclosure of the contract was necessary. He accepted that it might contain 'commercial or other sensitivities'.

Mr Forsdick said that there was no recognition in the defendants' case of the distinction between peaceful protest and direct action: 'They assume that if it's peaceful it's lawful, which is simply wrong. There is no right of direct action when it is unlawfully attempting to stop the Council doing its job'. But Mr Cooper responded that 'the Council is *not* doing its job. The issue here is not just the felling of healthy trees, but the unfair burden placed by PFI contracts on local authorities'.

Evidence

The first defendant to be sworn in was Cllr Alison Teal. The Independent Tree Panel (ITP) was, she said, not independent at all – it was set up merely to give the appearance that it was consulting with the community. The reality was that they 'just want to continue to assert their authority and do what they want to do'. The Council had been offered mediation but refused it. Ms Teal had had meetings with Council highways officers, including Paul Billington, in which he had talked at length about straight kerbs – 'they had to be perfectly straight with no deviation'.

Mr Forsdick said that Ms Teal had admitted to peacefully protesting tree-felling by standing inside barriers and to taking direct action, and had said she would go on doing so. She wanted the Council to change its policy and didn't like its democratic decision, something that she denied ('I very much like democracy'). Mr Forsdick referred to a photograph posted on Facebook showing her standing inside a safety zone; Mr Cooper asked her to clarify the background to the photo. It showed her standing beside an Amey arborist holding a mug of tea made for him by Ms Teal – a STAG mug. He had invited her behind the barriers to inspect a tree listed for felling.

As he took the oath, STAG founder Dave Dillner placed particular stress on the phrase '*nothing but the truth*'. Mr Dillner said that many councils had strategies in place to look after their trees and to ensure that felling was actually the last resort, not the first resort as appeared to be the case in Sheffield. He had never entered a safety zone where notices had been displayed, but on one occasion notices were attached to barriers after he had entered them. He was convinced that a target existed in the contract and the Council was trying to keep to it: 'My gut feeling is that the Council are determined to prove they are being strong and are not going to listen'.

Mr Dillner admitted to taking direct action and would 'continue to do so' but said that he would rather engage in dialogue with the Council. Asked by Mr Forsdick about a meeting in April 2017 with Cllr Bryan Lodge, Mr Dillner said that Cllr Lodge had inflamed the situation: 'He was banging the table and making it clear that it was the big stick and nothing else'. When he referred to the announcement by South Yorkshire Police that it would no longer prosecute tree protesters, Mr Forsdick remarked, to gasps from the galleries, that it was precisely because the police had reneged on its statutory duties that it was left for the Council to continue the process in the courts.

The third and last defendant to take the stand was Calvin Payne. He had been directly involved in the campaign since 2015, and had been arrested in November 2016 under the Trade Union Act. Mr Payne said that the number of ITP recommendations being overruled by Council decisions spoke for themselves. He referred to the residents' survey, with only one vote per household: 'That's not how democracy works in this country'. He believed in completely non-violent confrontation: 'To do otherwise would be pointless. Not one person [in the campaign] has committed an act of violence. We have a spotless record'.

Mr Payne described being served notices by Amey citing a law he'd never heard of: 'The Council and Amey are mixed up with one another – you can complain to the Council, but what you get is Amey'. Mr Forsdick referred to a statement in which Mr Payne said he wanted to force a change in Council policy, and noted that he was involved in direct action, or in suggesting it, almost every day. Asked if he had entered safety zones, Mr Payne replied that he had done so 'hundreds of times'.

The only other witness to be called was the Council's Director of Culture and Environment, Paul Billington, whose cross-examination stretched over two of the three days. Asked why an injunction was needed now, he said that it was only in the last twelve months that the council had been experiencing direct action, and the deadline for completing the Core Investment Period of highway works was 31st December 2017. In significant parts of the city there had been little trouble, but in some areas where STAG members were based there had been resistance. The number of incidents was increasing and many of them involved 'persons unknown'. This was causing Mr Billington a headache: 'The trees that are left are in the areas where it is most difficult to remove and replace'. He claimed to have had more than 1,000 meetings concerning the issue over the past two years: 'Nothing has taken up more time'.

Mr Cooper pointed out that the '6 Ds' was not a policy used anywhere outside of Sheffield, or recognised by any professional arborists, to which Mr Billington responded that this was 'a highway maintenance contract, not a tree programme. Arbs are not highway experts. The starting point is the impact on the highway'. Where trees were found to be damaging or discriminatory, tree maintenance and highway maintenance were intertwined, and Amey's work was therefore directly related to Section 41 of the Highways Act. He acknowledged that there was a section in the contract relating specifically to tree maintenance, and Mr Cooper pointed out that this had not been made available for the public to see. Mr Billington insisted: 'We fell trees as a last resort'.

He said that ITP advice came back to the Council but the 'decision of whether or not to fell the tree is made exclusively by myself'. Mr Cooper asked why 92% of ITP recommendations to save trees had been rejected. Mr Billington said that they involved 'engineering solutions outside the contract' that would have required 'additional resources not available to the Council'. Regarding the suggestion that Amey could have used a substance such as Flexipave to save trees, he said that he regarded tarmac as flexible paving. Mr Cooper pointed out that 'tarmac is only flexible when it's being laid'. Engineering solutions that were included in the contract had later been withdrawn as not viable.

Mr Billington said a 2007 survey of street trees by Elliott Consulting had indicated that 'there was a developing problem with the trees reaching maturity'. Mr Cooper said this was misleading and 'another example of cherry-picking – provided there are enough cherry trees around to pick from'. The survey had recommended the removal of only 1,000 trees, not half the tree stock. Mr Billington insisted that, 'Nobody on the Council has a penchant for felling trees – the goal is always retention of trees'. He recognised that some trees were regarded as more important than others, such as war memorial trees, the Chelsea Elm, and the Vernon Oak; in these cases the Council had 'tried to respond disproportionately and differently'. Mr Cooper suggested that they did not care about 'the swathes of rank-and-file trees. It's only if the tree is in the newspapers that you're interested. It's a political interest'.

At one point, to another audible intake of breath from the courtroom, Mr Cooper asked: 'Were you aware before you contracted with Amey that they were convicted of corporate manslaughter in September 2008?'. If the Council had known of such a conviction, or if it had not been disclosed at the contract negotiation stages but subsequently came to light, the contract should have been cancelled or renegotiated. But this proved a major miscalculation, as it was subsequently clarified that Amey had in fact only been convicted and fined under the Health and Safety at Work Act for failing to maintain adequate records, following a fatal accident.

Submissions

Making his closing submission, Mr Forsdick said that the Council's intent was to end the campaign of direct action, not to prevent peaceful protest outside safety zones, and to assert its right to fulfil its statutory obligations in the public interest: 'The defendants seek to prevent the lawful exercise of a statutory obligation. They assert that it is their right to engage in direct action in a democracy, and that is wrong'. He said that the issue was not whether the tree-felling process was justified, or whether the PFI contract was legal, but whether the protesters were entitled to take the law into their own hands.

The primary case against the defendants was the tort of trespass: 'Access to the public highway has to be for a reasonable purpose, and not to hold up public works. Being inside a safety zone is unlawful and therefore is by definition unreasonable – an illegal action cannot be reasonable'. The effect of the protests, Mr Forsdick said, was to destroy the right of the public to enjoy improved highways. Campaigners wanted to influence the Council's decisions but it was no longer at the decision-making stage: 'We've made our decisions and want to get on with putting them into practice. STAG thinks it has a veto over what the Council wants to do, and it hasn't'.

Mr Cooper submitted that the behaviour of the defendants was lawful because it was challenging a contract that was *un*lawful: 'Trees are being felled because of the domination of the contract by corporate interests'. By denying the public or the defence access to the contract, the Council was saying, in effect, 'Trust us'. It couldn't be said that the work had been significantly hindered, as nearly 6,000 trees had been cut down and only around 600 due to be felled were left: 'This is the final battlefield'.

He said that it was wrong to suggest that the defendants were 'cocking a snoop' (sic) at democracy: 'These are not professional protesters like Occupy London – it's a different genre. They are respected ordinary people who are speaking up for democracy. They are still ready to this day to negotiate. We submit that they are acting perfectly reasonably, perfectly proportionately. The defendants are argumentative, but there's nothing wrong with that – or where would we be in our profession?'.

At one point in the hearing the judge had asked about the relevance for the court of the question of the Council's engagement, or perceived lack of it, with campaigners, and Mr Forsdick had responded 'Nil'. Mr Cooper rejected that: 'This is the mood music that plays behind the case – Council bullying, Council intransigence, a refusal to sit down with people and do anything other than tell them what to do'. The Council's lack of engagement was what had brought the matter to its present pass: 'The reason we're here now is because the police wouldn't prosecute. Perhaps if it hadn't got to the point of the police being *asked* to prosecute we wouldn't have got to where we are now'.

Judgment

As the public record shows, the judge largely accepted the Council's submissions. In a judgment handed down on 15th August 2017, Justice Males granted the injunction, which took effect at one minute to midnight on 22nd August and lasted until 25th July 2018. The judgment pointedly expressed 'no view, one way or the other, as to the merits of the Council's tree-felling programme or the objectors' campaign. Those are social and environmental questions which are politically controversial and can only be resolved in a political forum'.

Justice Males accepted the evidence of Paul Billington, whom he described as 'an honest and reliable witness' and 'a conscientious and fair-minded council officer doing a challenging job', that 'the objective of the Council has been to retain trees where possible within the financial constraints under which the Council has had to operate. As he said, nobody in his team wants to fell healthy trees unnecessarily'.

The judgment also acknowledged the strength of support behind the tree campaign, as well as addressing the likely consequences of granting the injunction: 'Once the tree-felling programme is complete the exercise of any right to freedom of expression would necessarily have to be directed at protesting about what had been done, rather than campaigning to prevent it from happening. The trees would have gone'. The very effectiveness of the protests may have been one of the factors that led to the injunction being granted, as 'the location of the defendants' peaceful protest, within a safety zone, is an intrinsic part of the protest...the presence of the defendants within such a zone is itself intended to make, and does make, a powerful statement about what they see as the importance of the issue'.

However, the judgment concluded with a critical observation on one of the central planks of the defence's case: 'Although the defendants have made a number of sweeping allegations about the conduct of the Council and Amey, for example that healthy trees are being felled because Amey is exercising improper influence over the Council with a view to illegitimate profiteering, I have found that allegation to be detached from reality in the light of the evidence before me'.

Postscript (August 2019)

In the two years (at time of writing) since the trial a number of developments have taken place, more than can adequately be summarised here. They included the trials of several campaigners, including Alison Teal and Calvin Payne, for alleged breaches of the injunction, as well as a further court case to renew and extend it. But a few other things are worth mentioning.

Following the submission of Freedom of Information requests by members of the public and the intervention of the Information Commission, much more of the contract has been disclosed. The publication of previously redacted sections revealed that 17,500 street trees (out of a total stock of 36,000) were due for replacement under Streets Ahead. This fuelled suspicions that Amey and the Council were working to an arbitrary target, rather than felling trees according to need. After further protests attended by large numbers of both private security stewards employed by Amey and officers of South Yorkshire Police, a 'pause' in felling began in March 2018 and has continued to the present time.

Mediated talks between campaigners and the Council later in the year and a review by the Council of some of its previous decisions led to a number of trees being taken off the felling list and others having plans for their removal postponed. In future all candidates for felling were to be subject to 'joint investigations' by both Amey staff and highway engineers appointed by STAG, beginning in January 2019. The overwhelming majority of these investigations to date, observed and recorded by STAG volunteers, has confirmed that most of the remaining at-risk trees can be saved by the application of simple engineering solutions already provided for in the contract. The conclusion can easily be drawn that, had these solutions been applied earlier, hundreds or even thousands of trees could also have been saved without extra cost to the taxpayer. The question of why they were not applied remains to be answered.

HIGH COURT, SHEFFIELD, 27th October 2017

This one was all about the barriers. When is a safety zone not a safety zone? That was the question.

The trial at Sheffield Combined Courts was to hear Sheffield City Council's application to commit three protesters to prison for alleged breaches of the injunction issued in July: Councillor Alison Teal, Calvin Payne and Siobhan O'Malley.

David Forsdick QC, opening for SCC, described direct action as criminally unlawful. A small number of persons unknown, he said, were wearing masks to impede identification. Efforts to identify them were continuing, and further applications for committal were likely to follow.

Paul Powlesland, acting for Calvin Payne, raised a procedural matter: 'Which application notice are we fighting against?'. Since first making their application, the Council's lawyers had amended the details to include new allegations. Mr Forsdick explained that the first application for committal had been made on 6th October, but the incident involving Siobhan O'Malley was on 10th October. They had attempted unsuccessfully to serve notice on her before putting the documents in the post. Ms O'Malley had received them on Friday 13th October, leaving a one-day shortfall in the customary two weeks allowed to prepare a defence. Mr Forsdick nevertheless considered that this should have been sufficient time to seek legal aid. A friend speaking on Ms O'Malley's behalf said that she had not been able to do so, and that she was classed as a vulnerable person under the Mental Health Act.

Mr Justice Males, presiding (as he had over the injunction hearing), decided not to proceed with Ms O'Malley's case without her having had legal advice: 'The consequences could be very serious', he told her, 'the Council is asking me to send you to prison'. There would be a further hearing the following Friday 3rd November, by which time she could get legal representation. Justice Males advised her: 'If I were to hear you had been breaching the injunction before then that would count badly against you. So stay away from any safety zones or anything that looks like a safety zone'.

The allegations against Alison Teal and Calvin Payne concerned incidents taking place on Dunkeld Road on 25th September involving Mr Payne, and on Kenwood Road on 25th, 28th and 29th September, concerning them both. They were each accused of either entering a safety zone after one had been constructed or remaining in one after being asked to leave. In one instance, said Mr Forsdick, they were not sure whether Cllr Teal had actually entered a safety zone, but she had remained in one after it was erected. He referred to a photograph that appeared to show Cllr Teal in a safety zone on 28th September. She claimed that she was standing on private land on the grounds of Kenwood Hall Hotel, and although the photo showed that her feet were on cobbles Mr Forsdick could not definitely say that this was a patch of the highway maintained at public expense, so this allegation was discontinued.

Catherine Casserley, for Alison Teal, said that the failings of the claimant in this case were many. Entering and remaining in a safety zone were set out as separate acts in the injunction. The first application was not personally served on Cllr Teal – it only reached her when she received a bundle of evidence on 24th October. The claimant had brought more evidence and more photographs which apparently were available at the time of the first application, so there was no reason not to have provided them earlier. Cllr Teal had not had fourteen days to consider the amended application, only three. Therefore only the original application should be considered at this trial, Ms Casserley argued, not the amended one.

Justice Males noted that in her evidence submitted to the court, Cllr Teal said there was no safety zone present because the barriers were never completed. Allegations 1-3 in the application referred to Cllr Teal and Mr Payne entering a safety zone. The claimant sought to add 'remaining in' in order to cover the possibility that the safety zone had been erected around someone already there. It was claimed by the defence that the amendment amounted to a new allegation, though they related to the same incident and would depend on the same evidence. He was not persuaded that there was any prejudice to the claimants and allowed the amended application to be heard.

The late application against Calvin Payne also related to messages posted on social media. Mr Forsdick referred to a Facebook comment of 6th October allegedly encouraging people to breach the injunction: 'What I would really like is for as many people as possible to break the injunction on Monday morning'. Mr Powlesland argued that the late amendment necessitated a different defence from the first application and sufficient time had not been allowed to prepare it. The judge considered that Mr Powlesland had not explained his case sufficiently and that his objections were 'technical in the extreme'. The amended application was allowed.

The first witness to be called was Paul Billington, Director of Culture and Environment for the Council. Mr Powlesland questioned him about notices on barriers and warning letters handed out to members of the public, bearing an interpretation of the injunction different from that detailed in the court order. The injunction stated: 'For the avoidance of doubt a safety zone is that area delineated by barriers erected on the public highway around a tree to be felled'. The wording was important, said Mr Powlesland – standing inside barriers erected on a public highway was different from standing in a private garden or a public park. Why had SCC circulated letters which changed 'on public highways' to 'around any street'? Mr Billington: 'I can't answer that'.

Mr Powlesland said that the Council's wording could refer to gardens or parks. Mr Billington replied that they hadn't brought anyone to court who had been protesting in gardens or parks. 'But have you *threatened* to bring committal proceedings for standing in parks or gardens?' The council officer confirmed that documents had been handed out to people standing in parks, though not in gardens. He said that the Council's letter was within the spirit of the injunction. Mr Powlesland submitted that the wording of the letter was a threat, and SCC were wrong in issuing warning letters to people not standing within safety zones, and who were not in breach of the injunction. Mr Billington said that Mr Powlesland was 'suggesting there was a willy-nilly attitude on the part of the Council' in the handing out of the letters. The officer did not regard them as threats, though he could 'imagine bringing people to court is quite threatening'. Would Mr Billington agree that the letter and its wording were an abuse of the injunction? 'I believe we're right in bringing forward people who are clearly in breach of the injunction. Where there is no evidence we won't bring people forward'.

Mr Powlesland said that it was the defence's case that the Council had not come before the court with clean hands. He told the judge that 'it has distorted your words' in ways that are 'at best inaccurate, at worst misleading'. His client would say that this had affected his approach to protesting – SCC had repeatedly made threats of committal against people who were doing perfectly legitimate things.

Ms Casserley – who was so quietly spoken that those of us sitting on the press bench struggled to hear her at times – asked: 'The definition of a safety zone – what is it?'. Mr Billington said that it was the area around a felling, including safety barriers, but also 'structural barriers' such as a wall, as was the case on Kenwood Road. Justice Males interjected that, 'the definition of what constitutes a safety zone is a matter for me, not a witness, to determine'.

The next witness was Jason Wignell, an arborist working for Acorn Environmental Management Group, which was sub-contracted by Amey to carry out the fellings on these dates. Under cross-examination by Ms Casserley, Mr Wignell explained the procedure of setting up barriers around a safety zone. On 25th September at Kenwood Road he had become aware of Alison Teal's presence on site, but could not say at what time she had arrived. Barriers had been set up around three sides of the tree to be felled, extending to a second tree; on the fourth side was a wall standing on private property, against which had been positioned some barriers, albeit with gaps between them. Mr Wignell considered that a wall or ledge could be used as the front line of a safety zone, irrespective of whether the wall was on private or publicly-owned land.

Ms Casserley referred to written evidence by Amey's Director of Account, Darren Butt, and photographs taken by him and Mr Wignell, which showed that where Cllr Teal was standing, there was no barrier present. Justice Males interjected, referring to a similar view in another photo apparently taken at the same time, where neither the wall nor barriers were visible. Mr Wignell argued that it didn't matter whether a barrier was in place as the fourth side was the wall.

Mr Powlesland said that at the time Mr Wignell had arrived on site Calvin Payne was not in a safety zone. He asked if at one point in the erection of the barriers Mr Payne's conduct changed from being lawful to unlawful. Mr Wignell said that he '*intended* to step inside the barriers', and believed Mr Payne's intention was to be unlawful from the beginning – he intended to stop the felling. Mr Powlesland noted that the wording of the injunction did not mention intentions.

Mr Wignell said that a safety zone 'is the area around the tree' – there can be four or fifty barriers so long as it's complete. I don't believe a break in the barriers is an incomplete safety zone'. Asked if he would consider it a breach of the injunction for someone standing in private property to prevent the felling of a tree, Mr Wignell said that he would. Mr Powlesland said that none of the photos submitted in evidence showed a completed work zone. Mr Wignell said that a gap had briefly been left to allow a lady to leave her property. He said that he could see Mr Payne was in the safety zone and referred to a photo in which he could clearly be seen standing on a public highway. Ms Casserley referred to later photos showing a very large gap in the barriers and Mr Forsdick pointed out the lack of barriers along the wall.

Darren Butt was sworn in and affirmed his written evidence. He had arrived on site at Kenwood Road at 11:30 on 25th September. Ms Casserley asked if Cllr Teal could have been there all along and the barriers put up around her; Mr Butt said yes: 'I told her that everyone there was in breach of the injunction. There were barriers leaning against the wall with gaps between them, but that was irrelevant because the wall itself formed a safety zone'. Ms Casserley asked if he had said, 'You're in enough trouble already' and Mr Butt agreed that he had 'made a flippant remark to that effect'. He said that while Cllr Teal was sitting on the wall, she wasn't in breach of the injunction. The barriers against the wall were 'belt and braces – to prevent people from reaching the wall'. Ms Casserley asked if he gave crews any instructions about where to put up barriers. Mr Butt said that it depended on the situation on the day: 'I don't consider it necessary to put barriers along the wall'.

Mr Powlesland said that it was important that notices put up on barriers accurately reflected the judge's actual order: 'The notice on the barrier was inaccurate, wasn't it?'. Mr Butt said that he was just carrying out his job and was 'not a legal specialist like yourself'. Mr Powlesland asked if he would agree that a wall or hedge can be a natural barrier and Mr Butt replied in the affirmative. Mr Powlesland suggested that this was not consistent with the injunction, which specified a safety barrier: 'Where is the boundary line, the barrier or the wall?'. Mr Butt: 'It depends on the gap between the barrier and the wall'. Mr Powlesland referred to a photo of a person between a barrier and a wall: 'Is he in the work zone? It matters, as someone may go to prison'. Justice Males intervened to note that it was not an appropriate question to ask a witness if someone deserves to go to prison.

Mr Butt argued that the edge of the highway was the edge of the work zone, irrespective of where the barriers are. Mr Powlesland asked he had directly witnessed Mr Payne entering the safety zone on 29th September. Mr Butt replied that he had not, but he was already in the safety zone and was asked on numerous occasions to leave it: 'I told him he was in the wrong place, and you can see from his Facebook post that he knew he was in the wrong place'.

Justice Males observed that the letter on the barrier in the photos looked like it came from SCC/Amey and not from him. He pointed out to Mr Forsdick that the issue with Cllr Teal was that she was not standing in an area completely ringed by barriers. He reiterated the wording in the injunction: 'Barriers erected on a public highway'. Mr Forsdick submitted that when a safety zone is intended to go up to a natural boundary like a wall then the wall serves as the barrier. Barriers were put up to prevent people from getting onto the wall without going through the safety zone. There was no authority on this, but he thought it was understood.

The judge demurred: 'I'm afraid I'm against you, and the case against Ms Teal is dismissed'. The decision had been made so summarily that listeners in the courtroom were startled, not sure that they had heard correctly. Mr Forsdick attempted to recover the situation, referring to barriers having been moved around and Cllr Teal's admission that she had been in the safety zone. Justice Males stood firm. He accepted her written evidence that she had left the work once it was completed and agreed with the defence's argument that a safety zone as specified in the injunction needed to include barriers erected on a public highway forming a complete, unbroken perimeter on all sides; 'natural barriers' like walls did not count. He could not be sure, to a criminal standard, that Cllr Teal had been inside a completed safety zone in these terms, so the case against her had to be dismissed. The judge noted of Calvin Payne, however: 'The case against Mr Klein [sic!] is another matter'.

Paul Brooke was then called as a witness for the defence's argument of the Council's 'unclean hands'. He talked about being filmed by evidence gatherers outside Amey's depot at Olive Grove and being confronted by them reading injunction letters with omitted words or changed wording. Nothing about the injunction seemed to show that the protest at the depot was illegal. Amey workers had been telling people that standing in private gardens to protest fellings was illegal. He believed that the injunction was being interpreted in this way because SCC's intentions were being frustrated. Notices were being served on people whose cars were legally parked or who were standing on private land or park land. On Meersbrook Park Road barriers were set up around people in the park and an evidence gatherer said they would be in breach of the injunction. Mr Brooke received the notice as a threat – he was certain that he wasn't in breach but other people present were not so some left the area.

Mr Forsdick referred to the case against Mr Payne on Dunkeld Road on 25th September. He was in the safety zone and knew from his conversation with Mr Billington that he was in contempt of court. Counsel referred to Mr Wignell's evidence and a tweet in which Mr Payne admitted breaching the injunction, as well as the Facebook post and photographic evidence. Mr Payne had admitted breaking the injunction and evidence showed that he was ignoring the authority of the court.

Mr Powlesland pointed out that in the photos there were no barriers visible delineating the work area: 'There is doubt that barriers were erected all the way around the work zone'. Justice Males observed: 'You're getting onto Ms Teal's band wagon'. Mr Powlesland said that the claimant had not fulfilled the burden of proof that the zone was clearly delineated and Mr Payne could not be found in contempt as a result. He asked which version of the injunction they were looking at, the one SCC/Amey wanted or the one His Lordship had made. Justice Males observed: 'The only one I'm interested in is the one that I made. The serious protesters were alive to the exact wording and fully understood the implications. Why should I conclude Mr Payne was any different?'. He added that it was not only SCC that had an interest in the injunction being obeyed but also the court: 'There's a public interest in the rule of law'.

Justice Males said there was no doubt that Mr Payne was in contempt of court on three counts, including the Facebook post. He concluded the hearing by saying that judgment would be given the following Friday, 3rd

November, after he had had time to reflect. He said that in the case of Siobhan O'Malley the Council might want to reflect that she had learned her lesson – this seemed to many present to be a heavy hint to the Council to drop the case against her.

Speaking outside the courtroom to the assembled media (viz: your humble scribe and a trainee reporter from the *Guardian*), Mr Billington welcomed the judge's decision on Mr Payne: 'A clear finding that Mr Payne was in contempt of court will mean that people will respect the injunction'. He said that the case was about who makes the law in Sheffield – it could not be people on the streets. He expressed concern about protesters wearing masks, which he called 'a sinister and cynical move'. The Council would press on with the work after their success with Mr Payne.

HIGH COURT, SHEFFIELD, 3rd November 2017

Although there had been a lively demonstration outside the court in support of the defendant, inside the courtroom it was very quiet, without much chatter to disturb the mood of sombre expectancy. Everyone assembled there was anxious to hear Calvin Payne's fate and all feared the worst, himself included.

Neither Paul Billington nor David Forsdick was present. Instead of the expensive QC, Yaaser Vanderman opened for the claimant, outlining the sentencing options open to the judge. In pleading mitigation, Paul Powlesland emphasised the unusual context of the case, reiterating his previous point about the Council's 'unclean hands' in its treatment of the injunction's wording. Justice Males interrupted, asking him to reflect on whether this line of argument was helping his client. Continuing, Mr Powlesland said that Mr Payne was a man of good character and displayed a level of civic engagement that was laudable. He had not acted for personal gain but as part of a group joined by many who believed passionately and deeply in their cause, and had campaigned for it in many different ways, as far as possible within the law. The judge observed that 'nothing in the injunction prevents him from doing any of those things'.

There followed a discussion over costs. Mr Vanderman said that the previous week's bill amounted to £62,000, and the present week's £32,000, after discounts. Mr Powlesland said that it was outrageous to spend so much on the proceedings – SCC had not explained why they had thought it necessary to retain an expensive QC when the case could have been handled by a junior barrister. He said that it smacked of overkill and a claimant who didn't seem to care about costs because they would be met by the taxpayer or by the other side. Mr Vanderman replied that the case was of great interest to the city and the country so it had been important to 'engage silk'.

After retiring for a brief recess, Justice Males returned with his judgment. Mr Payne was found in contempt of court in three respects: two of standing within safety zones and one of using social media to encourage others to break the injunction. A fourth allegation, concerning the Dunkeld Road incident, was not proven because photographic evidence appeared to show that the safety zone included a private garden hedge on one side where there were no barriers visible. Thus the judge could not be sure to a criminal standard that it was a completed safety zone even if Mr Payne's own words suggested otherwise.

In sentencing, Justice Males said: 'Imprisonment is in general reserved for cases where there is serious and contumacious flouting of an order of the court'. He continued that it was clear from their deliberateness, their repetition and his declared intention on social media to continue to break the injunction and encouragement of others to do likewise that Mr Payne's actions were indeed 'a serious and contumacious flouting of the court's order'. (Note: for the avoidance of doubt, contumacious means 'stubbornly or wilfully disobedient to authority'. I looked it up so you don't have to.) Mr Payne's prominence as a campaigner meant that others were likely to be influenced by his example, and be tempted to breach the injunction also. The claim offered in mitigation by the defence that the Council had acted against the letter of the court order was not held to be relevant.

At this point, however, the judge stepped back from the brink. As Mr Payne was 'a man of good character and essentially, I believe, a law-abiding and decent man', who had not committed further breaches since the application for committal was served, he was to be given one final chance. A sentence of three months' imprisonment suspended for one year was passed for each of the three counts, to run concurrently. Mr Payne was also ordered to pay the Council's legal costs in his case, assessed as £16,000. Justice Males added, 'I urge you to think hard about what I have said. It is up to you now'.

Following the hearing, Mr Payne told this reporter that he thought the judge's words and his sentence were fair in the circumstances. He said: 'I won't stop what I do, but I will change what I do'. He continued: 'Everything we do is because of something the Council does. There are definitely circumstances in which I wouldn't have broken the injunction. We spent several months trying to find ways around it. We'll follow the injunction – I hope the Council will as well'.

SHEFFIELD MAGISTRATES' COURT, 24th May 2018

The hearing began with the defendant, a feather in his hair, asking for a drink of water. The court usher obliged, but couldn't fit the cup through the gaps between the glass screens surrounding the dock. There followed a little unrehearsed choreography as the magistrate asked the defendant ('Mr Compin') to stand further away from the bench and to remove his hands from his pockets ('I'm sure you don't mean anything by it').

There were three allegations, all concerning events taking place on Abbeydale Park Rise, Dore, on 5th March: two cases of assault, one involving a knee to the groin and the other the throwing of two punches; and thirdly, the wilful obstruction of a police officer in the course of his duty.

In relation to the alleged assaults, the defendant admitted (through his solicitor, Mr Ben Manovitch) that physical contact had been made, but claimed that it was justified because he was defending himself from attack:

he was being manhandled by several men larger than himself and pain was being inflicted on him through his fingers being pulled back. He denied obstructing or intending to obstruct a police officer – a charge based on his refusal to give his name when asked, on the grounds that his identity was already known.

The prosecution will claim that the security staff the defendant allegedly assaulted were seeking to prevent him committing a criminal offence under Section 303 of the Highways Act – interference in lawful highway works. The defence will challenge the legality of the particular tree-felling and will deny that the defendant was committing, or intending to commit, a criminal act. They will claim that security staff therefore had no legal right to restrain him, because no crime was in the process of taking place.

At the conclusion of the brief hearing the magistrate set aside a full day for the trial on 3rd August (beginning at 10:00am at the Magistrates' Court). The existing bail condition which prevents the defendant from attending the sites of tree-fellings is henceforth lifted, but the condition preventing his making contact with the two alleged victims remains in force.

Asked for a statement afterwards, Mr Compin produced a pack of Bruce Lee Playing Cards ® and requested this reporter to pick a card, any card. The resulting randomly-chosen statement read as follows: 'Be proficient in your field as well as in harmony among fellow men'. The defendant affirmed that he did indeed intend to be proficient as a poet and to live in harmony among his fellows.

This may not be a straightforward trial.

HIGH COURT, SHEFFIELD, 5th June 2018

It was a day of two gasps, bracketed by a pair of curve balls.

The trial of Paul Brooke, Benoit Compin, Simon Crump and Fran Grace began with a lengthy preamble about procedural technicalities. Also discussed was the question of whether the testimony of two expert witnesses brought by the defence counted as either expert or relevant. The judge ruled that, for the time being at least, it did not.

But after more than an hour of tedious formalities Justice Males gave his first showstopper. 'Judges, like everyone else, occasionally read the papers', he said, noting the present 'moratorium, for whatever reason', on tree-felling. Seeking the committal to prison of four Sheffield citizens was a serious matter. Did it have the backing of democratically-elected councillors, and not just unelected council officers?

Yaaser Vanderman (acting for Sheffield City Council) seemed a little thrown. He said that the ultimate decision to proceed lay with the Council's legal office, not its leader, though lawyers would have taken into account the views of Council members. This was not the decisive response the judge was asking for. Abruptly, he rose and stated that he was not prepared to proceed 'until you can give me a clear answer, yes or no'.

This unexpected intermission gave rise to much speculation among the STAG supporters gathered in the public gallery. Who would take responsibility for the decision to prosecute? Was there indeed a 'separation of powers' as recently claimed by Council cabinet members?

Phone calls having been made backstage, the court resumed. Mr Vanderman reported that SCC's legal Officer, Steve Eccleston, had spoken with Council Leader Julie Dore the previous day, when 'she was happy for the proceedings to go ahead'. A further conversation had just confirmed that 'she positively agrees that proceedings should be brought'. So that made clear where the buck stops.

Mr Vanderman then introduced twenty-five minutes of video evidence. Technical presentation was regrettably lacking in finesse. A first attempt to run the footage from a USB stick resulted in a frozen screen and muffled sound. After much mucking about with laptops, the video clips (stretched to a CinemaScope shape, with consequent optical distortion) were eventually played from CDs. No awards for showmanship here.

The evidence pertained to events on Meersbrook Park Road (MPR) in December and January (four separate occasions) and on Abbeydale Park Rise (APR) in March. It was claimed that the videos showed the defendants inside safety zones around trees and refusing to leave immediately when asked. This, it was stated by Mr Vanderman, put them in breach of the injunction.

Seven witnesses were called to testify in support of the prosecution. Six took the stand today: arborist Jason Wignell, evidence gatherer Jake Webb, security guards Ross Henderson, Ricky Learman, and Richard Milligan, and Streets Ahead account director for Amey, Darren Butt. (Council officer Paul Billington will be heard from tomorrow.)

Several of the witnesses were asked to clarify claims made in their statements that one of the defendants, Simon Crump, had remained inside a safety zone for ten minutes when video footage in fact accounted for only three and a half minutes.

Under cross-examination by Paul Powlesland, acting for Simon Crump and Fran Grace, Acorn arb Jason Wignell was asked to define what considerations determined the size of a safety zone. The Meersbrook Park Road zone on 18th December was narrower on the park side, where it abutted the railings, than on the highway side. At one point it was extended further on the road side, with the defendants still inside it. The zone was then reduced again after they had left the zone. Why had it been extended, and why had the reasons not been explained to Dr Crump and Ms Grace at the time? Mr Powlesland argued that the extension was done not for safety reasons, but to force protesters out of the area.

Owen Greenhall, acting for Paul Brooke and Benoit Compin, related another incident on MPR, when the police had facilitated a protest. Mr Wignell had been in his vehicle, frustrated at being unable to move, and Mr Greenhall said that he had spat water at protesters through his window. Mr Wignell's response – 'That's not true' – drew a loud intake of breath from the public gallery.

Evidence gatherer Jack Webb disputed Benoit Compin's claim that he had not known the terms of the injunction before 10th January, because he had previously seen Mr Compin at 'half a dozen' felling sites; he was recognisable by his French accent and guitar. But Mr Greenhall pointed out that in his affidavit Mr Webb had not said that he had recognised him, only that he had been 'informed by his colleagues' that the protester was Mr Compin. So how could he recall how many times Mr Compin had attended fellings before, or say what he knew about the injunction?

Mr Greenhall pointed out to Mr Wignell and Mr Webb that when Mr Compin entered a safety zone on MPR, for a total of three and a half minutes, another protester was already inside and had already brought work to a halt. The protester was still there when Mr Compin left the zone. Mr Webb was asked: 'So when you say, "As a result of protesters' actions the tree was not felled that day", you don't mean that *Mr Compin* prevented the work.'

There was a long, long pause – around a minute – before the reply: 'I still believe his actions were the reason why the tree was not felled that day'.

Mr Greenhall played a video clip several times in slow motion, showing what appeared to be an uppercut from a security guard to a protester on MPR. Mr Webb denied that the video clearly showed a punch being thrown. Mr Greenhall also referred to a woman protester being knocked to the ground and dragged across the floor: 'I put it to you that the force that was used on that protester was not reasonable force'. Justice Males asked if this was a fair question to put to a witness. Mr Greenhall responded: 'It's what he says in his statement – that reasonable force was used'.

Darren Butt asserted that every tree is assessed on its own merits and that the tree on MPR was due to be felled because it had dislodged a kerbstone; installing a thinner kerb would be only a temporary solution as the tree would continue to grow and dislodge the kerbstone again. Mr Greenhall referred Mr Butt to a photo of the kerb: 'Would you agree that the displacement of the kerb is minor?'. Mr Butt: 'I accept that the kerb is displaced'. Mr Greenhall: 'Would you accept it is minor?'. Mr Butt: 'I accept it is displaced'.

Mr Butt had personally asked Dr Crump and two other campaigners to leave an incomplete safety zone on MPR. Mr Powlesland asked if anything else in the zone had prevented the felling that day, and noted that a car was parked next to the tree: would the felling have gone ahead with the car still inside the zone? Mr Butt said that parking services would have been called to remove it: 'We don't chop trees over cars'. Another audible gasp from the public gallery led the judge to say that it was unfair to have witnesses' statements commented on in this way. A security officer in the gallery took up position alongside the front row, presumably ready to eject anyone who breached decorum.

Mr Powlesland showed Mr Butt a video shot from under a tree while it was being felled and asked if Mr Butt considered the work safe. Mr Butt said that he did, having previously investigated the video and been complimented on the work by professional associates. He did not believe the branch being cut was directly over the videographer as it appeared.

Mr Butt said that he was aware that some complaints of unreasonable force had been made against security guards. Several witnesses were asked about what they considered reasonable force; Justice Males noted that they can say 'what was done and why it was done, but whether it was reasonable is a matter for me'. He referred to a video taken on MPR on 22nd January in which a woman was heard to cry out in pain. The judge stated that it was not a security guard who had caused the woman to cry out, but an unidentified woman on the park side who had grabbed her hand through the railings. And on that dramatic note, he rose and the court was adjourned.

The trial continues tomorrow and Thursday.

HIGH COURT, SHEFFIELD, 6th June 2018

It's official: geckoing is not, repeat not, a legal loophole. Justice Males made the point emphatically, implicitly agreeing with STAG co-chair and defendant Paul Brooke that it is a legitimate way for tree campaigners to protest without breaching the injunction.

The judge's remark was made in response to prosecuting counsel Yaaser Vanderman, towards the end of the second day of hearings in the trial of Mr Brooke, Benoit Compin, Simon Crump and Fran Grace, for alleged breaches of the injunction forbidding the intrusion of protesters into safety zones surrounding tree works.

Owen Greenhall, acting for Mr Brooke and Mr Compin, continued the cross-examination of security guard Richard Milligan, who agreed that the principal function of security staff was the physical removal of protesters from the area of a tree-felling, rather than the infliction of pain: 'You said that if a person was crying out in pain the first response would be to check that they were OK'. The aim was to use minimal force and to reduce the risk of injuries.

In his written affidavit Mr Milligan had mentioned guards using a slow rocking motion to extract protesters rather than a sharp jerk, which had a greater risk of injury. He agreed that if someone was carried out acting as a 'dead weight' then it would be appropriate for four guards each to take hold of a limb. 'But you wouldn't punch anyone?' – 'No'. Mr Greenhall then showed the video clip, previously played yesterday, which appeared to show a security guard throwing a punch during a fracas. Mr Milligan had not been involved but had been standing beside the evidence gatherer shooting the video. He said that he could not be sure that the clip showed a punch being thrown.

Mr Greenhall said that one female protester on Meersbrook Park Road (MPR) had been removed from park railings twice on 22nd January. On the second occasion, her cries of pain had led bystanders including Paul Brooke to break through the barriers surrounding the safety zone. Mr Greenhall showed a harrowing video clip of three security guards attempting to remove the woman, one by apparently lifting her up with his shoulder. A voice could be heard shouting to police off camera: 'Officers, you're witnessing an assault!'

At this point, Justice Males interrupted proceedings to relay a report passed on to him from a barrister not involved in the case, who had witnessed the apparent intimidation of council officer Paul Billington in the waiting area outside the courtroom. The judge said that the intimidation of a witness about to give evidence was a serious matter and could itself be considered a contempt of court. A short recess was called to deal with this. When the hearing reconvened, Mr Vanderman stated that Mr Billington had not been aware of the report and that the alleged intimidation was nothing more than he was used to. Nevertheless, the Judge stated that if the offence was repeated he was prepared to exclude the public from the building; he didn't want to do this as he believed in an open court, but the press were sure to report the matter.

Resuming his cross-examination of Mr Milligan, Mr Greenhall showed further video footage of the removal of the female protester from the MPR railings. He pointed out that the clip showed sharp tugs being used, not a rocking motion as Mr Milligan had described in his affidavit. Mr Greenhall also noted that several other security staff, including a female guard, had been standing nearby as the protester was carried away; surely they could have been employed to pick her up bodily rather than dragging her? Mr Milligan replied that he would have said they were placing her on the ground rather than dragging her.

It was then the turn of Paul Billington to take the stand. Before admitting him to the courtroom, the judge (no doubt recalling previous hearings at which Mr Billington had testified and the experience of other witnesses the day before, as well as today's report of alleged intimidation) reminded the public gallery that there must be no audible reactions to his evidence.

Anyone who felt unable to comply with this should leave.

Mr Vanderman referred to the decision to fell a tree on Abbeydale Park Rise (APR) that was displacing the kerb by approximately 30mm. Mr Billington said that this was not a decision he had taken (it had been made by his predecessor, Simon Green) but one that he supported. The particular species of tree (wild cherry woodstock) had an expected rate of growth of 10mm per year, so it was likely that any remedial work done on the kerb would have to be repeated in a year or two's time. Replacing the tree with a similar species but with a slower rate of growth would give the chance of a tree with a long-term future.

Paul Powlesland, acting for Simon Crump and Fran Grace, asked Mr Billington for his definition of a highway safety zone and whether it could be erected for any reason other than to provide a safe working area. Mr Billington said that SCC and Amey's mistaken understanding of what constituted a safety zone (that it did not include natural boundaries such as walls and park railings) had been cleared up at a previous hearing and that since then they had been adhering to the definition specified in the injunction.

Mr Powlesland drew Mr Billington's attention to a photograph of a warning sign displayed on MPR and asked him several times if he would agree that it gave an incorrect version of the injunction wording: yes or no? (Justice Males interjected to ask him to keep his voice down.) Mr Billington said that he did not recognise the sign, and was not qualified to say whether its message was within the terms of the injunction: 'If I said yes or no I'd be guessing'.'

Mr Powlesland asked if Mr Billington agreed that when Amey staff say people are in breach of the injunction, protesters were within their rights to challenge them over their confusion regarding the injunction. Mr Billington said that he was not aware of any confusion. There was some confusion over whether it was tree workers or tree campaigners who were confused. Mr Powlesland said that it was Simon Crump's case that he had been confused about the extension of a safety zone. Mr Billington said that Dr Crump had been involved in the campaign from the beginning and so 'the chances of him being confused about anything are amongst the lowest of anything in this case'.

Mr Powlesland referred to a matter discussed in a previous hearing in October, in which he had cross-examined Mr Billington over the contents of an erroneously worded letter distributed to a number of campaigners standing in Meersbrook Park, outside of safety zones, and whether it could be construed as threatening. Mr Powlesland promised not to go there again, drawing a wry smile from the council officer, but noted that it was an example of why campaigners such as Dr Crump and Fran Grace could not take the word of SCC, Amey or Acorn staff, because of inconsistencies in information previously provided by them.

Mr Powlesland then referred to Mr Billington's statement in the original injunction hearing that only 6,000 trees were to be felled. Mr Billington insisted that he had been referring only to the Core Investment period, and that previous as well as subsequent Council statements and press releases had made that clear. He had never meant to suggest that 6,000 was the total number to be felled in the entire contract term.

A slide was presented showing clause 6.38 in the Streets Ahead contract, specifying 17,500 trees to be felled. Mr Billington said that he could see how the contract could be interpreted to imply a target of that figure but that it was not the case. Bids for the contract had been made on the basis that up to that many trees could be replaced if necessary, without additional cost to the Council. This was standard contacting procedure. But no discussions had taken place regarding any such target: 'It doesn't exist'.

Mr Billington did not know, and could not say, how many trees would be felled over the course of the contract, but 10,000 was the Council's 'best guess'. He said that it could be argued that it might be in Amey's best financial interests to replace less than the maximum figure and that at some point the Council might want to discuss that – 'but that's in twenty years' time'. Mr Powlesland said that citizens of Sheffield wanted to know how such decisions were being made now. The Council had not released its annual Tree Management Programmes – Mr Billington said he hoped they would be released.

Mr Greenhall wanted to discuss the decision made to replace one particular tree, situated between house numbers 22 and 24 on Abbeydale Park Rise. The Independent Tree Panel advice for this tree was as follows: 'Remove two kerbs and replace with droppers. Narrow footway by 150mm over 15m. Remove and relay 15m of stone edging. Proceed with engineering solution and associated work, if funds are available'. However the Council's final judgment was to remove and replace the tree.

Mr Greenhall said that the displaced kerb was not inconveniencing members of the public. Mr Billington noted that the Streets Ahead contract specifies no undue deviation from a straight kerb line, a principle he upheld: 'As a cyclist, I wouldn't want any deviation'. He said that he spent a lot of time cycling close to kerbs and wouldn't want his name against a decision that could cost a human life.

Mr Greenhall said that the ITP had proposed engineering solutions to retain twelve healthy trees on APR, and that the same decision had been made for all of them. Following a 'tree walk' on APR with Amey representatives, a resident had asked why so many trees were being felled and was told that there was no 'appetite' at SCC for the use of engineering solutions and that the Council preferred replacement. Mr Billington said that instead of 'appetite' he would substitute 'ability', as the ITP recommendations involved, 'at least in part', bespoke engineering solutions sitting outside the contract that would require additional funding that was not available: 'If an exception was made in the case of one tree you would find yourself making them in many cases'.

Mr Greenhall said that a mistake had been made in the decision-making process in consideration of this particular proposal. Mr Billington: 'I reject that'. Mr Billington said, 'This is a highway renewal contract, not a patch-and-mend contract'.

Defence

Opening the case for the defence, Mr Powlesland called Simon Crump to the witness stand. He corrected a typographical error in his affidavit, in which a photograph of a warning sign had been mistakenly identified as having been taken on 15th January rather than 16th January.

Mr Vanderman said that Dr Crump was a committed campaigner.

Dr Crump: 'I would say more of a keen amateur'.

Mr Vanderman: 'You say "committed" in your written evidence'.

Dr Crump: 'Oh dear'.

Mr Vanderman: 'You have sat under countless trees and have attempted to frustrate tree-felling'.

Dr Crump: 'I don't accept that. I've attempted to delay it'.

Mr Vanderman: 'Until... what? Someone else comes along?'

Dr Crump: 'No. I want to delay felling so the two parties can get together and negotiate a sensible solution'.

Mr Vanderman: 'You intend to go on entering safety zones?'.

Dr Crump: 'Absolutely not'.

Mr Vanderman referred to a message Dr Crump had posted on the STAG Facebook page last week in which he said that he did intend to do so. Dr Crump drew the court's attention to the inverted commas he had placed around 'safety zones': 'I would caution you not to get excited about it and remind you that I have a PhD in English literature. I choose my words and my punctuation very carefully. I used inverted commas to refer to "fake" safety zones. Mr Vanderman noted that the post had subsequently been removed and suggested that Dr Crump had deleted it because he knew it would affect the case. Dr Crump could not be sure whether it was he or the page moderators who had deleted it.

Mr Vanderman referred to the incident involving Dr Crump and Fran Grace on MPR on 18th December. Dr Crump: 'It felt to me that we were standing outside of the safety zone and no-one explained to us why we were not. It felt like entrapment – they were trying to get us on something'. Mr Vanderman said that a trap allows entry, but not exit. Justice Males said that there were two senses of 'trap': to be stopped from leaving or to be tricked into breaching the injunction. Which sense did Dr Crump mean? 'The second one'.

Mr Vanderman showed video footage of the incident and said that Dr Crump and Ms Grace had been asked to leave a total of eight times. Dr Crump said that he did not count them – the experience had been quite stressful. Mr Vanderman: 'You could have left the disputed area at any point during those five minutes'. He pointed out that Dr Crump had been talking to Calvin Payne on the other side of the fence, who had not been 'drawn in' to the area because 'he knew a safety zone was being erected'. Dr Crump said that if Mr Vanderman wanted to know Mr Payne's reasons for leaving he could ask him, and pointed to the public gallery where Mr Payne was sitting. Justice Males said if the court wanted to hear from Mr Payne he could be called as a witness and upheld Mr Powlesland's objection to Mr Vanderman's speculation.

Dr Crump said that he and other campaigners had been lied to continually by SCC and Amey: 'They should get the wording of the injunction straight and stick to it'. Dr Crump said he had been genuinely confused as to what was happening: 'It was the first time they had tried this particular tactic. I also believed it was not a proper safety zone – you can hear me saying that.' He had been 'confused by the lies they have told us over weeks and months – they will tell us anything'.

Mr Vanderman: 'Not once did you ask the question why the safety zone was being enlarged'.

Dr Crump: 'I was hoping they would tell me'.

Mr Vanderman referred to the later incident on MPR: 'I was standing near the railing with my arm around the railing, some distance from the tree'. Mr Vanderman said that Darren Butt had asked him to leave: 'Because it wasn't a complete safety zone, I refused'. Two other campaigners who had linked arms with Dr Crump decided to leave the zone but he stayed. He denied having prevented the erection of barriers, which could have been placed in front of him as they had been 'on many previous occasions'. Mr Vanderman asked why he was holding onto the railings: 'Because people are removed, sometimes violently'. This had happened the same day further down the street.

Mr Vanderman: 'You knew they wouldn't fell the tree with you standing there'.

Dr Crump: 'No, there was a car there. It was part of the deal'.

Justice Males referred to the undertaking that Dr Crump had signed not to participate in direct action. Dr Crump: 'Which is very similar to the injunction. I signed it in good faith and have taken it seriously. As far as I am concerned I have not broken the injunction'.

Fran Grace next took the stand. Mr Vanderman pointed out that she is a former primary-school teacher and asked if she had any qualifications in arboriculture or Health and Safety. She had not. Ms Grace has been involved in the campaign for about three years but stated in her written affidavit that she had become more involved after the Rustlings Road incident in November 2016. What did 'more involved' mean?

Ms Grace: 'I decided not to be an armchair activist and to become more active. I felt a lot more strongly about the issue and that I wanted to be involved'.

Mr Vanderman: 'Including attending felling sites?'.

Ms Grace: 'Yes'.

Mr Vanderman suggested to Ms Grace that the reason she stayed with Dr Crump in the 'disputed area' on MPR was because she was 'upset'.

Ms Grace: 'That's not what I said. Several trees were in danger of being felled. I did what I usually do: take photos, talk to other campaigners, be a witness. I felt trapped because I had no intention of being inside a barrier – they built the barriers around me. I don't know why they did that'.

Mr Vanderman: 'Is it your defence that the reason for the safety zone being extended was to trap you or to get you to move away?'.

Ms Grace: 'Entrapment'. Ms Grace accepted that she and Dr Crump had been asked to leave eight times.

Mr Vanderman: 'The reason you stayed was because you knew they wouldn't fell the trees with you standing there'.

Ms Grace: 'That's not the reason. I was genuinely confused about why they had to enlarge the safety zone when there was already a safety zone. I didn't know if there was a legitimate reason for building the barriers around us – it didn't make sense. I was confused and bewildered. I hadn't gone in, they had put me in'.

Mr Vanderman: 'I'm putting it to you that you wanted to prevent the tree being felled'.

Ms Grace: 'If I'd wanted to prevent the tree being felled I'd have taken direct action. But I didn't, I just stood there'.

Mr Vanderman: 'You didn't ask why the barriers were put up?'.

Ms Grace: 'No, but I'm not sure if I'd have believed what they said anyway.'.

Justice Males interjected: 'Was the level of distrust such that you would not have believed anything they said?'.

Ms Grace: 'That's precisely the point. The level of distrust between us was already so high'.

Mr Vanderman: 'You asked about the wording of the injunction?'.

Ms Grace: 'I was asking about the legitimacy of what they were doing – whether there was a valid reason for extending the safety zone'.

Ms Grace said that on the day she had not heard what one of the security guards had said to her, as played back on a video of the incident: 'I don't hear very well'. (Ms Grace had worn a hearing loop throughout the two-day hearing, including during her time on the stand, and asked for a number of Mr Vanderman's questions to be repeated.)

In the video clip, Ms Grace was heard asking to see 'the 'real' injunction'. Mr Vanderman asked if she knew what the injunction said: 'Yes, but I also knew that there had been at least one other injunction posted that was incorrect'.

Paul Brooke was called to the witness stand. As he passed Ms Grace they touched hands briefly in a gesture of... solidarity? Affirming his written affidavit, he asked for a single word in his statement to be corrected: 'intention' should have been 'attention'.

Mr Greenhall asked Mr Brooke approximately how many other felling sites he had attended besides the one in this case: 'I can't be precise – in the region of approximately forty occasions'.

Mr Greenhall: 'Have you ever prevented a tree from being felled?'.

Mr Brooke: 'Not to my knowledge'. He had not entered a safety zone since the granting of the injunction, after signing an undertaking.

Mr Greenhall referred to an incident in which an arborist had allegedly spat at him. Mr Brooke confirmed that this was Jason Wignell, who had been called as a witness for the prosecution the previous day: 'I made a complaint to the police and have not heard that the case is closed'. He had not been aware that Mr Wignell was to be called to give evidence before yesterday.

Mr Brooke said that on the day of the incident in question, 22nd January, he had a brief conversation with Calvin Payne after arriving on Meersbrook Park Road, in which he was told that a protester had been beaten up. This was the incident in the video appearing to show an uppercut. Mr Greenhall played a video of a masked female protester being prised away from railings by security guards and screaming in pain. Mr Brooke had not witnessed this incident directly but later was shown the video clip by another campaigner, Alan Simpson, on his telephone.

Mr Greenhall: 'Did you notice anything?'.

Mr Brooke (with voice breaking slightly): 'I noticed a woman being assaulted'.

Another video clip was shown of the same female protester being pulled away from railings a second time, with the sound of screams and cries of 'Don't hurt her!'.

Mr Brooke said that during this incident he was standing about eight metres away to one side and about a metre or two behind the Heras fencing: 'I heard the woman scream and thought at the time that the woman was being assaulted again. At one point a security guard stood on or kicked the woman. I assumed she was in pain and being assaulted again'.

A lengthy video clip was shown depicting the fencing collapsing from the weight of the campaigners behind it, including Mr Brooke pushing it, followed by a surge of people into the safety zone, again including Mr Brooke, who swore at a security guard. After a moment he retreated back outside of the zone, kicking the now fallen fencing. The video continued for some minutes, showing other protesters chanting and forming a protective circle around a threatened tree.

Mr Greenhall asked Mr Brooke his intentions: 'To distract their attention and prevent the woman from being assaulted'.

Mr Vanderman said that Mr Brooke is a resident of MPR and walks past these trees every day: 'It's natural that these trees are particularly dear to you'. Referring to the incidents shown on the videos, he asked if Mr Brooke was aware of any complaints being made by the victim to the police. Mr Brooke said that she could not complain because she did not want to identify herself. Others had made complaints on her behalf but they had not been taken seriously by police because they had not been made by the victim.

When Mr Vanderman asked if he knew who the victim was, Mr Brooke addressed the judge, saying that he would prefer not to incriminate another person. Justice Males said that this was his right, though it might affect how he assessed Mr Brooke's evidence. When asked again by Mr Vanderman, Mr Brooke said that he did know the victim; Mr Vanderman then asked her name, and he declined to answer. Justice Males queried the point of asking and Mr Brooke again insisted that he would not wish potentially to incriminate someone.

Mr Vanderman asked why Mr Brooke had not asked police officers on the scene to intervene. Mr Brooke said that they were too far away, and earlier in the day they had told him that they were only there to observe. He noted that since the occasion under discussion, police had changed their tactics and at subsequent protests observed more closely, standing within a few feet of protester removals. Mr Vanderman said that he knew police officers did not consider that unreasonable force was being used on 22nd January. Mr Brooke: 'I knew they weren't acting'.

Mr Brooke denied that he considered 'any' use of force was unreasonable and said that what he thought about the actions of security staff in general was immaterial: 'I believed a woman was being assaulted at that time'.

Mr Vanderman suggested that he was feeling angry during the incident. Mr Brooke said that he had felt a mixture of emotions, including anger, and that it had been 'a traumatic experience for me – it was very unlike me'. Mr Vanderman said that he had not entered the safety zone because he believed a woman was being assaulted but because he was angry at a tree being felled on his street. Mr Brooke: 'Which tree, this one or the fifth?' He said that he had been at fifteen or twenty fellings and had not reacted in this same way.

Mr Vanderman had referred earlier in the proceedings to 'geckoing' as an 'alleged loophole used to prevent or delay felling without breaching the injunction'. Mr Powlesland asked Mr Brooke about 'geckoing'. Mr Brooke: 'It is not a loophole! People stand by a wall or railing so that the barriers have to be erected in front of them. Nine times out of ten, that is what happens. Geckoing is not about exploiting a loophole, it's about asserting a legal right of protest'. Justice Males asked why it was called 'geckoing', and on being told the answer said, 'It's ironic, then?'.

Mr Brooke said that arborists would not specify what size a safety zone has to be because they keep changing it as it suits them – 'One metre, two metres. Not once to my knowledge have they ever approached a householder to ask permission or taken them through the courts to cut branches oversailing their property. What they will do is turn up at five in the morning when no-one is looking'.

Benoit Compin has elected not to testify and will stand by his written affidavit. The full video, lasting around four minutes, was played of his performance of a poem inside a safety zone where work had already ceased. Security guards are shown asking him to leave, and he argues and tries to continue before agreeing to leave.

Mr Vanderman said that he had wanted to cross-examine Mr Compin about contradictions in his statement but would instead make a list of questions to pass on to his barrister, Mr Greenhall.

Submissions

Commencing his closing submissions, Mr Vanderman stated that a 'safety zone' was not defined as per the injunction in the undertaking that Simon Crump had signed because the injunction had been issued after the undertaking was signed. He argued that in his case the 'natural meaning' of the term should apply, which included the notion that a safety zone extended to boundary walls and railings.

Concerning the incident on 18th December, Mr Vanderman said there could be no dispute that Dr Crump had remained in a safety zone, as he was within barriers erected around a tree. He said that this is sufficient to show that he and Fran Grace were in breach of the injunction. They could not have been trapped if they had been given the option to leave, which they could have done at any time.

Mr Vanderman referred to the evidence given by security guard Ricky Learman, that there is a default practice of setting up a safety zone around a tree to be felled before the lead arborist determined how big it needed to be. He said that arborist Jason Wignell's evidence confirmed that a safety zone was only ever extended for safety purposes, including the size of the tree and the number of staff and vehicles involved. Mr Vanderman said that Dr Crump and Ms Grace had not asked why the zone was being extended, which would have been the obvious thing to do.

Mr Vanderman stated that Dr Crump had also said the zone was not complete. Justice Males corrected him: he is heard in the video evidence saying that it was 'not a proper zone' but he does not explain in the video why it was not.

Mr Vanderman stated that Dr Crump did not want to leave the area in which he was standing because he knew felling could not take place with him there. He accepted that he had been standing there to delay the felling. In the incident on 16th January, Mr Vanderman said that there could be no question that he prevented safety barriers being erected flush against the railings. It could not be to prevent him from being removed because there were no security staff there at the time. The video of him standing there alone lasted around an hour and a half in total (Justice Males said that he did not want to have to sit through a video of someone standing by railings for an hour and a half). The size of the tree meant that the barriers had to be up against the railings, according to Darren Butt's evidence, and this was a matter of expert opinion Dr Crump was not qualified to speak about. The only possible reason for Dr Crump to be standing there was to delay the felling.

Justice Males asked what limit there was on where Amey can put a safety zone: 'Is there a good-faith test or a need for a reasonable belief or do they have to be right?'. Mr Vanderman stated that it was a matter of rationality.

Mr Vanderman said that, to the extent that there were any doubts as to Dr Crump's intentions, they were dismissed by his Facebook post. It was very relevant that he had deleted it: 'He says that he can't remember whether it was deleted by administrators or himself, but it was only last week'.

Justice Males said that it would be a bit odd if the regime was different for those who had given an undertaking and those who had not – it would be disorderly if some people could be in one place and not others. He said that it could be argued that the definition of a safety zone in the injunction did not apply to the undertaking because the injunction didn't exist when the undertakings were signed.

Mr Vanderman said that Fran Grace attended felling sites two or three times a week, so she clearly knew what the injunction said: 'I put it to her that geckoing was a loophole'. Justice Males interrupted to say that, 'It's not a loophole, is it?' It was a question of what the injunction did or did not allow. Mr Vanderman said Ms Grace was asked to read an injunction notice but accepted that she may not have heard the instruction.

Mr Vanderman said that Benoit Compin accepted that he had entered a safety zone on 10th January and remained for three and a half minutes.

The trial continues tomorrow.

HIGH COURT, SHEFFIELD, 7th June 2018

You will by now know the verdicts in three of the four cases brought before the High Court in Sheffield this past week. But for the record, I conclude herewith my first-hand account of the trial with this summary of the third and final day of the hearings.

Disclaimer: This day's proceedings are the most challenging to cover because they depended substantially on intricate legal argument, in which I cannot pretend to be an expert. My handwritten transcript will undoubtedly have oversimplified, skated over or omitted important details from the counsels' submissions, especially in their references to arcane case law which was not readily accessible to those of us in the public gallery. Nevertheless, I hope to have got across the gist in layman's terms of at least some of the key issues at stake. Corrections from other observers in the courtroom would be welcome.

Before continuing his final submissions begun on Wednesday, Yaaser Vanderman (acting for Sheffield City Council) raised several preliminary matters, including the possible disclosure to the judge of Facebook posts which he claimed showed that Benoit Compin knew of the injunction before 10th January. Mr Compin accepted that he breached the injunction by entering a safety zone for a total of three and a half minutes on Meersbrook Park Road (MPR) on that day. Counsel for the defence objected that such disclosure might prove unduly prejudicial, but Mr Justice Males stated that he would like to see all relevant materials before sentencing.

In a second incident, on Abbeydale Park Rise (APR) on 5th March, Mr Compin had been within a safety zone from at least 3:02pm to 3:50pm. He had attempted to prevent the removal from the zone of a female protester and had then climbed a tree, where he remained for some time despite the pleas of police officers to come down. He is heard on video evidence saying, 'I'm going to breach the injunction again'. This Mr Vanderman described as 'a brazen admission of defiance'. The incident took place after Mr Compin had been notified that he had breached the injunction previously, and after the sentencing of Calvin Payne and Alastair Wright for contempt.

The defence had argued that the particular tree on APR should not have been marked for felling. But Mr Vanderman said that the defence's reading of Section 41 of the Highways Act was too narrow. He referred to the previous judgment of Mr Justice Gilbart that the lawful authority (in this case, SCC) has the right to remove trees as it sees fit. The tree was being removed to maintain the highway: 'We say that's the end of it'. Mr Vanderman said that the defence had not submitted evidence that 30mm was not a significant deviation from the kerb line. The question of whether alternative engineering solutions could have been used to retain the tree was irrelevant to the injunction. There was no record of Mr Compin having previously objected to the felling of the tree – the proper approach would have been to approach the Council.

Justice Males noted that further growth of the tree potentially disrupting the highway in future was not a concern that had been recorded in the ITP report or the Council's final decisions. There was apparently no significant further growth in 2018 compared to 2016, when the decision to fell had been made. (He remarked that this might not be relevant to his judgment in the trial.)

Mr Vanderman then turned to the case of Paul Brooke, whom he described as a 'long-time campaigner' and 'one of the major protagonists' in the tree dispute. Mr Brooke was one of those who had signed an undertaking before the granting of the injunction and had later given evidence at the trial of Calvin Payne. His defence is that he entered a safety zone by breaking through a barrier because he believed that a masked woman protester was being assaulted by security staff.

Referring to the video evidence of the forcible removal of a masked female protester from railings on MPR, to which Mr Brooke said he was reacting, Mr Vanderman said that nothing in the clips was especially shocking. He said that Mr Brooke had entered the safety zone in an aggressive and violent manner. As a result of his actions, tree-felling had been suspended for a number of weeks.

Mr Vanderman referred to Section 3 of the Criminal Law Act 1967, which states that, 'A person may use such force as is reasonable in the circumstances in the prevention of crime'. He said that security staff employed by Amey had been given extensive training and briefing in the use of reasonable force in this context. He argued that the methods and the amount of force used by them were reasonable, as shown by the fact that the removal of the female protester took only thirty seconds. As she was acting as a 'dead weight' when being carried away, security staff acted reasonably in rotating her and placing her down on the floor.

Mr Brooke's defence was also based on Section 3. Mr Vanderman said that there had never been a civil case in which Section 3 had been entered or accepted as a defence. He noted there were some 'novel' legal points in play but that, 'On all legal authority Mr Brooke's defence must fail'. There were a number of police officers in close proximity who could have acted if they thought a crime was being committed by the security guards. Their decision not to do so showed that they did not believe an assault was taking place. This, said Mr Vanderman, was 'fatal' to Mr Brooke's defence: 'There can be no defence of a breach of an injunction if the act was deliberately undertaken. If you do the act, that's it'.

Justice Males observed that the police officers in attendance were not especially close to the incident of the removal and did not appear to be looking in its direction. Mr Vanderman said that people had been complaining to the police so it could not be said that they were unaware of what was going on; they were close enough to the safety zone. He said that Mr Brooke did not deny that the police officers considered the force reasonable but that he proceeded to enter the safety zone anyway. Mr Brooke did not approach the female protester to ask if she was OK; other people may have been in the way, but he did not even try. The judge noted that a security guard had done a 'good job' in calming Mr Brooke down and persuading him to leave the safety zone.

Mr Vanderman said that Mr Brooke's action of lifting up the barrier was not the breach of the injunction, but rather his entering the safety zone. He should have let the police officers present do their job and not take the law into his own hands. Once the incident started he didn't approach a single police officer; even if their backs were turned to the safety zone, they were still there. There was no sufficient nexus between entering the safety zone and the action of preventing a crime; at the time Mr Brooke entered, the woman had already been removed from the railings and was on the floor.

Justice Males said that video evidence appeared to show that the potential assault to which Mr Brooke said he was reacting was over by the time he entered the zone. Would it be a defence that he had a mistaken belief in the possibility of an assault? The judge remarked that it would be 'a bit harsh' if the law said that someone had to stand by 'and watch someone being beaten to a pulp'.

Defence

Opening his submissions for the defence of Simon Crump and Fran Grace, Paul Powlesland said that this case was not just being fought in the courts but in the Council chamber and on the streets. Despite the assurance given to the court on Tuesday that SCC took responsibility for bringing the case, outside court it had said otherwise in order to avoid the political consequences. The question of who had authorised the prosecution of four campaigners had been raised in Council by Dave Dillner on Wednesday. Mr Powlesland quoted Council Leader Julie Dore's response and said that it proved that the decision to proceed was not taken by councillors but by council officers, even if she agreed with that decision.

At this point Justice Males interrupted and asked Mr Powlesland to move on. He was satisfied with the assurance he had been given by the Council. Mr Powlesland asked, 'But who is the Council?'. The judge responded that if he didn't know that he should look it up.

Mr Powlesland returned instead to the question of what constitutes a safety zone. The definition contained in the injunction is that it was an area surrounded on all four sides by barriers in order to create a safe working area. This latter part was crucial to the definition: the zone must have been created for this purpose. It could not just be barriers around a tree.

Mr Powlesland noted that Dr Crump had signed an undertaking to the court but Ms Grace had not. He recalled a previous discussion in which it was noted that the definition of a 'safety zone' specified in the injunction did not exist when Dr Crump and others had signed their undertaking, because the court order for the injunction had not yet been given. Mr Vanderman had argued that, in the case of those who had signed the undertaking, the 'natural definition' of a safety zone, one in which 'natural barriers' such as walls and railings counted as one side of the zone, should therefore apply.

Justice Males noted that it would be 'disorderly' if two protesters standing side by side were under a different regime and were subject to different conditions, because one had signed an undertaking and the other had not. The Council was not proceeding against defendants on the basis that they had breached an undertaking but that they had breached the injunction, even though they were not 'persons unknown' but rather 'persons known'.

Mr Powlesland said that on 16th January Dr Crump had been holding onto a railing in the gap between two barriers. The prosecution claimed that he had thereby prevented the erection of a complete safety zone. Mr Powlesland said that if Mr Vanderman's previous suggestion about the distinction between two definitions of 'safety zone' were to be applied, Dr Crump could not be said to have prevented the erection of a safety zone under the definitions of the injunction, because he was already in one under the alternative definition. The judge asked incredulously: 'Are you suggesting that Dr Crump was in breach of his undertaking?'.

Mr Powlesland said that it has often been the case that protesters have stood between railings and barriers and work has still continued. This could have been done on this occasion. Amey workers have shown that they are willing to sacrifice standards to fell trees over the heads of protesters. If it had been anyone but Dr Crump standing beside the railings, they would have done that.

Mr Powlesland said there was no clear evidence to explain why the safety zone had been enlarged in the incident on 18th December. Arborist Jason Wignell's evidence had been very confused, and had given different explanations for the necessary size of safety zones. Mr Wignell had stated that a risk assessment was done for both the initial zone on MPR and the extended one, but a copy of neither risk assessment had been submitted in evidence. The lead arborist on 18th December, Dom Barrett, has not come forward to give evidence.

The result was that the defendants, Dr Crump and Ms Grace, had been confused. There had been previous occasions when Dr Crump and others had been told they were in breach of the injunction when in fact they were not. It was therefore reasonable for them to question the information they were given, and to assume that the safety zone was not being enlarged for *bona fide* reasons. The bad faith that existed between the parties meant that rather than leave the zone when they were asked, they wanted more information. They wanted to pursue lawful, peaceful protest but this situation was one they had not encountered before.

There was constant flip-flopping from Amey about what constituted a safety zone and what protesters are allowed to do. Very wide powers had been given to a private company to exclude citizens from parts of their own city. People need to know in advance what they are and are not allowed to do, in order to undertake the most effective form of legal protest. Given the high standards of legal proof required, Mr Powlesland submitted that Dr Crump and Ms Grace were not in breach of the injunction.

Owen Greenhall, opening his submission for the defence of Paul Brooke, said that his submission was based on Section 3 of the Criminal Law Act 1967, which allowed citizens to use reasonable force to prevent a crime being committed, including the defence of another party. He said that the wording was clear and had to apply to alleged breaches of a court injunction.

Mr Greenhall explained that duress should be defined by the urgency to act; 'urgent action' is close to the classic definition of self-defence.

Mr Greenhall said that it could be argued that following the use of reasonable force by Mr Brooke in pushing over a safety barrier in the course of defending another, the safety zone itself as defined by the injunction ceased to exist. He said that a belief that an assault was taking place should be both honest and reasonable, but that they were usually run together and that it would be odd to separate them.

At this point, proceedings were interrupted when Mr Greenhall's mobile phone rang. Justice Males observed drily: 'There are judges who would say *that* was a contempt of court!'. Continuing, Mr Greenhall said that in criminal law it would be wrong to send someone to prison for using reasonable force so long as that person acted under a reasonable conviction. The sanctions for contempt were likely to be higher than for criminal proceedings.

He said it was not necessary for someone in Mr Brooke's position to reach a correct conclusion about what was happening – two people in the same position might have reached different conclusions – but it was enough to show that he had reached a reasonable conclusion in the circumstances. Mr Brooke himself might have done something different on a different occasion. Mr Greenhall said that if the judge did not agree that this was a valid defence in law then he would like to submit it in mitigation.

Moving from legal argument to the facts in Mr Brooke's case, Mr Greenhall said that it was clear the defendant had a genuine belief that the female protester was under attack – the strongest evidence for this was the video footage, in which Mr Brooke was plainly and manifestly concerned. With the information he had received that day that security guards were manhandling protesters and the video footage he had seen which supported that information, Mr Brooke had formed the opinion that security staff were prepared to cause significant pain to female protesters. What was relevant was not whether the security guards were in fact using reasonable force but what Mr Brooke perceived and the opinion he formed, and that it was a genuine and reasonable belief that a woman was being assaulted.

Mr Greenhall said there was no evidence in the video footage to show that security guards made any effort to check that the protester on the ground was OK. It looked to Mr Brooke like she had been assaulted, and was about to be assaulted further by a guard standing on her or kicking her. As soon as Mr Brooke realised she was OK, he left the safety zone. He did not join other protesters who formed a ring around the threatened tree.

Mr Greenhall said that Mr Brooke is not someone who advocates direct action, as proven by the fact that he has not acted similarly on other occasions. Nor is he abusive towards security staff – as shown in the video evidence, they know him by name. As to the police, Mr Brooke had been told earlier in the day by officers that they were taking a hands-off approach. It does not follow that because of their failure to act they didn't think an assault was taking place. It was their inaction that prompted him to act. He conveyed that he was not happy with the situation by his facial expression seen in the video footage and his use of profanities in a manner that was out of character. Justice Males commented: 'I think he succeeded in conveying that he was not happy'.

Turning to the case of Benoit Compin, Mr Greenhall said that he would concentrate on the legal argument for the incident on APR as Mr Compin had admitted breaching the injunction on MPR. He argued that in the case of the particular tree Mr Compin was defending, Section 41 of the Highways Act (concerning the Council's duty to maintain the highways) was not engaged. Contrary to SCC's reasons for choosing to fell the tree, the engineering solutions needed to retain it were covered in the PFI contract. A deviation from the kerb line of approximately an inch was not significant – Mr Greenhall said there must be hundreds or thousands of instances of kerbstones with similar deviations around the city where trees are not marked for felling.

Mr Greenhall said there was no evidence of any sustained examination of the rate of change of the tree growth. There was a question therefore of whether the works around this tree should be included within the injunction. In a criminal case a defendant had the right to query an administrative mistake that had led him to court; this principle should apply to a civil case as well.

Responding to Mr Greenhall's submissions, Yaaser Vanderman said that the key issue with Paul Brooke was his state of mind. Mr Brooke accepted that he was far away from the incident with the female protester and could not see the detail of events, but Mr Vanderman said that contrary to his evidence his state of mind was one of 'blind indifference' to what was happening to the woman. Mr Vanderman described the notion that all four police officers on duty turned their backs to the incident as 'unreal': the whole purpose of their being there was to observe.

Judgment

Justice Males said, in his characteristically understated manner, that he found the cases against Dr Crump, Ms Grace and Mr Compin to be proved. He reserved judgment on Mr Brooke until a later date because there were points of law that he needed to consider.

Mr Vanderman read out Facebook posts from Mr Compin, referring to posts from Calvin Payne and Alastair Wright which demonstrated his knowledge of the injunction and willingness to break it. He also showed the judge a photo of Mr Compin sticking up two fingers, indicating his attitude of defiance.

The defence moved to pleas in mitigation. Mr Powlesland said of Ms Grace that there was no evidence of any other breaches of the injunction or of attempted breaches. She had attempted lawful peaceful protest on other occasions. This breach was not deliberate or planned – she came prepared to stand outside of a safety zone as she had previously done. She remained within the safety zone for between three and ten minutes – a very short time indeed. This had little to no impact on delaying a felling or the tree-felling programme. She was acting on the advice of Dr Crump and was clearly not a ringleader or organiser. Whether she was aware of being in breach should have an impact on sentencing. She was an honest and genuine witness – she was evidently confused in the video footage, was asking for clarity and got none. This put her at the low end of the spectrum of seriousness.

Mr Powlesland said that much the same applied to Dr Crump. He did not plan to breach the injunction by entering a safety zone but found a zone built around him. In the second incident, Dr Crump aimed to delay a felling without breaching the injunction. He believed 'geckoing' to be a lawful, generally accepted way of protesting without breaching the injunction. With reference to Dr Crump's Facebook post, the wording and punctuation were crucial – on previous occasions he had been correct in not being in breach of the injunction when told that he was. If the judge was minded to give a custodial sentence then a suspended sentence would be appropriate – Dr Crump was a man of previous good character.

Mr Greenhall submitted that any sentence for Benoit Compin should also be suspended. In the first incident he was inside a safety zone for a total of three and a half minutes and caused minimal or no delay to the felling. Mr Compin accepted that he was aware of the injunction, though not of its specific terms. He had been in France at the time he made his first Facebook post in October. Mr Greenhall noted that people on Facebook communicated with like-minded others and sometimes made statements which went beyond their actual intentions.

In the second incident, Mr Compin believed he was standing up for people in general. It was not a pre-planned incident – what prompted his actions was concern for an elderly female protester with multiple sclerosis. Nothing in the judgment should take into account Mr Compin's actions towards security guards, which are the subject of a separate legal hearing. In the wake of this he was upset and used language he was not proud of. Mr Compin was aware of how his actions had brought him to the court, and he would not do anything to bring him there again.

There was then a short recess of around twenty minutes before the judge returned to pass sentence on the three defendants found guilty of contempt. The full judgment has been posted, so there is no need to report it in detail here. In summary, Simon Crump and Benoit Compin were each sentenced to two months' imprisonment, suspended for one year. In Fran Grace's case, the fact of being found in contempt was held to be sufficient punishment. All three were persons of good character. Justice Males concluded by thanking those in the public gallery for their quiet and respectful attention to the proceedings.

After the judge rose and left the courtroom, observers in the public gallery stood up to applaud the defendants down below. Outside the courtroom in the waiting area and again outside the court building, other supporters applauded and hugged them as they emerged.

Technically, the three defendants had lost. Morally, it felt like a victory.

Postscript (January 2022)

In a judgment given remotely on 21st June (Mr Justice Males not being present in Sheffield at the time), the case against Paul Brooke was dismissed. A moral victory had become a legal one.

SHEFFIELD HIGH COURT, 11th July 2018

Undramatic scenes punctuated the inaction at today's tedious non-hearing at Sheffield High Court. Boredom ran high as anti-climax followed breathlessly on anti-climax and non-event piled on non-event before culminating in irresolution.

No arguments and no evidence were submitted in court by Sheffield City Council, as it sought to extend and vary the current injunction against direct action at tree-felling sites. Instead, legal teams representing both the council and tree campaigners spent most of the day huddled in meeting rooms as they tried to thrash out a deal behind closed doors.

The hearing was due to start at 10:00am sharp. The four defendants challenging the new injunction application, along with three other non-signatories plus SCC's Paul Billington and Amey's Darren Butt, were accordingly assembled with their respective legal representatives in court at that time. But nothing happened. Of the judge there was no sign.

Animated conversations broke out among the lawyers as they pointed to laptop screens and scribbled on notepads. Others present gathered in groups or pairs to chat idly while issues were discussed on their behalf. (All this was of course inaudible to public onlookers up in the gallery.) Then the various participants began disappearing to off-stage consultation rooms, until by 11:30am the courtroom was empty apart from its staff.

It was not until 1:00pm that everyone reconvened and Mr Justice Robinson took the bench, only to call a lunch break after ten minutes of procedure. Further waiting around followed the intermission, and it was exactly 4:23pm when the judge was called again. He dropped a strong hint that both parties should agree a 'consensual compromise' before the court reconvenes in the morning. And that was it.

Campaign supporters gathered in the public gallery were forced to speculate on what might lie behind SCC's failure to make its case in open court.

Were the Council's legal advisers unsure of the legitimacy of their bid (available for scrutiny in published documents) to restrict civil liberties to the point where residents would be forced to produce ID to gain admittance to their own homes, and so to stifle peaceful protest that the only valid opposition would take the form of private fantasy?

Or were they seeking to press home their proposed prohibitions on such freedoms as slow-walking and to enforce their mania to include 'natural boundaries' like walls and railings within the definition of safety zones?

Perhaps we will find out tomorrow. Meanwhile, observers salvaged what stimulation they could from today's non-event:

Suspense as Dave Dillner tried to complete his newspaper crossword.

Spectacle as Paul Brooke performed a spontaneous dance for the gallery.

Colour from defence counsel Paul Powlesland's snazzily-patterned socks.

Intrigue as Paul Billington and Darren Butt read, apparently for the first time, the 'We Want You to Complain [about Amey]' leaflet.

Shock at the revelation that SCC's counsel Katharine Holland QC is being paid £35,000 for a day and a half's work.

The case continues. It might even get started.

SHEFFIELD HIGH COURT, 12th July 2018

New readers start here:

On 21st June, thirteen tree campaigners were sent letters by Sheffield City Council, asking them to sign an undertaking based on a reworked version of the injunction issued last summer and due to expire on 25th July, which SCC hoped to renew and extend for three years. They were invited to sign with a promise that if they did so they would face no legal costs. They had a week to decide how to respond. All refused to sign.

There were nine named defendants in the revised injunction application, four of whom actively opposed it in court while the other five merely declined to sign. The remaining four had been sent the injunction papers but were not officially named as defendants. SCC repeatedly failed to respond to requests from the campaigners' legal teams to clarify the status of these others, deemed 'persons unknown'.

Got that? Good.

The second day of hearings began with a private meeting of the two legal teams in offices at the Town Hall. If the entire negotiations had taken place there, the taxpayer could have been saved a bill estimated at £75,000.

Instead, a lengthy process of what co-defendant Paul Brooke described as 'obscene horse-trading' took place on court premises – but mostly not in the courtroom. By my count, proceedings before the judge occupied barely an hour across the two days. Only Sheffield City Council can say why it refused campaigners' repeated requests to settle the matter beforehand and instead employed two London barristers at huge public expense which it will not get back.

The Ealing-comedy air of the situation was beautifully illustrated at the pre-hearings rally, when two of the defence lawyers pulled up on a pushbike. No taxis or limousines for Paul Powlesland and Ben Manovitch – a bicycle built for one, but ridden by two, was good enough.

Inside the courtroom, the waiting began again. The minutes and hours ticked by even more slowly than yesterday, when at least observers had been excited to know what was going on and indulged in much speculation. Now we knew, and it wasn't exciting at all.

Some incidental highlights of the day:

- Conscientious and fair-minded council officer Paul Billington spilled water over his desk and had to mop it up with tissues. The spillage left an unsightly stain on the desk covering.

- Mr Powlesland set hearts fluttering by flashing the waistband of his knickers at the public gallery. I don't remember this happening in *Rumpole of the Bailey*.

- Conscientious and fair-minded council officer Paul Billington ate a green apple (as indeed he also had yesterday), disposing of the sticky label out of sight below his desk.

- The unexpected sound of a knock – usually the signal to rise for the judge – caused everyone in the courtroom to snap into position in anticipation of His Honour's appearance. It was a false alarm and we all relaxed.

- Conscientious and fair-minded council officer Paul Billington was given a talking-to by a security guard about his use of a mobile phone inside the courtroom. Muted applause from the public gallery.

His Honour Judge Graham Robinson was first called at 11:02am. Katharine Holland QC, representing SCC, opened by pointing out a formatting error in the layout of the draft injunction, which could have caused a misunderstanding. There followed a discussion about the three categories of defendants: the four who had actively defended, the other named five and 'persons unknown'.

In his opening submission yesterday, Owen Greenhall, representing three of the defendants, requested that no order be given for costs due to the amounts involved. This was accepted, so that each side met its own legal costs. These were considerably greater for the Council than the campaigners.

The business of the court largely having been settled off-stage, Justice Robinson asked Mr Greenhall and Mr Powlesland whether they wished to remain for the rest of the proceedings, as there was essentially nothing left for them to do; they elected to stay. The judge also identified those defendants present who had not agreed to sign it; Simon Crump reminded him that he should be referred to as Dr, not Mr, Crump, and the judge – a cheery, jovial sort of fellow – apologised for his error.

Ms Holland proposed to guide His Honour through her skeleton argument as it applied to the non-signatories and referred to the reading that had been provided for the judge. 'I've had plenty of time to do that, thank you!', he noted. There was no need to refer in detail to any part of the evidence that had been submitted, including the statements made by Darren Butt and Paul Billington, but Ms Holland said that it showed the 'continuing need for injunctive relief'.

Justice Robinson then stated that he was not going to 'draw any adverse inferences' about anyone who had refused to sign the new injunction, as it was 'substantially different' from the one originally awarded by Mr Justice Males last year. This is a point he would return to more emphatically in his closing statements later in the day.

Ms Holland said that the application for a revised injunction had come about because changing circumstances had shown that they were necessary to preserve the intentions of the original injunction. She itemised some of the changes that had been agreed to the first draft of the new injunction. These included the stipulation that slow-walking and other methods of delaying contractors could be used for up to twenty minutes at a time in any given area on any given day. The definition of a safety zone excluded action taking place on private property. Ms Holland also noted that the prohibition on 'encouraging' forbidden activities did not extend to general words of encouragement and support for the campaign posted on social media.

His Honour stated that there was no need for him to intervene in the form of the injunction that had been agreed and that he granted the injunctive relief sought. It was desirable to provide the injunction document as a single file – a combined order – rather than separate ones for the different categories of defendants. The court (which had been booked only for the morning session) was then adjourned for more than two hours while the revised injunction was typed up and printed out.

Proceedings resumed at 2:16pm, when His Honour brought out four copies of 'the draft' for the lawyers to proof read. Instructions given, he left again at 2:18pm. The two teams of lawyers then busied themselves checking the text, pens poised for corrections. The draft duly annotated and resubmitted, the Clerk of the Court returned at precisely 3:18pm with newly printed copies of the agreed 'final' version. But it proved not to be: defendant Rebecca Hammond noticed that her name had been omitted from the front page, while that of another, Graham Turnbull (not present in court), had been misspelled. The copies were collected in once again.

Justice Robinson took his seat for the last time at 3:25pm. He thanked everyone for the courteous way in which they had conducted themselves over the two days. He reiterated the point made previously by Justice Males: that the court expressed no view one way or the other as to the merits of the tree-felling programme or the tree campaign. He was there only to adjudicate, but in this case his role had been more like that of a mediator. He commended both parties for demonstrating what could be achieved through negotiation.

Most important was this closing observation: 'If I may say so, the defendants who did not take an active part were eminently justified in not signing the original undertaking'.

Council representatives had asked the defendants and their team if they could release a joint press statement. Campaigners agreed on condition that the statement stipulated that the pause in felling would continue until formal mediated talks had been arranged. SCC could not guarantee that, so there was no joint statement. Once again, the Council had snatched defeat from the jaws of victory.

Outside the court, co-defendant and STAG co-chair Paul Brooke instead made his own statement on behalf of the other defendants. Courtesy of Mr Brooke, I reproduce it below in full:

"Thirteen people were threatened with huge court costs to force them to sign an injunction. Thirteen people refused to be bullied. Collectively we opposed the application and faced £75K in costs. They applied for three years, with a ban on legitimate slow-walking and delaying felling crews. They wanted to curtail ordinary people's freedom of speech by preventing people posting encouraging comments on social media. They wanted to stop people standing on their own property to defend their trees. The outcome is that they spent £75K on horse-trading wording with us. We had asked them to do that for free before starting a court process. The new injunction makes it clear that anything that happens on private property is not restricted by the injunction. The Council agreed that we can slow-walk and delay felling every day. The term is not three years but eighteen months. The judge said the terms are a fair and proportionate balance between the competing interests of the parties. NONE of us have signed the undertaking but the four of us [who defended] will obey the injunction'.

AUTHOR'S NOTE

These accounts were based on notes taken at first hand by the author during the court hearings and in most cases written up immediately after for publication on the STAG website the following day. The major exception is the report on the trial of 27th October 2017 and subsequent sentencing, which was commissioned for this book and written up from contemporary notes, supplemented by the official judgment, in March 2021.

The aim of the reports was to provide a more detailed account than the necessarily abbreviated summaries that appeared in newspapers to serve as STAG's own record of the cases. I also wanted to capture something of the flavour of the hearings themselves for campaigners who could not personally be present in the courtroom. Their inclusion in this book will, I hope, help to complete the picture of the campaign provided by other witnesses.

Sheldon Hall, January 2022

DIARY OF A TREE PROTESTER

HEATHER RUSSELL

Heather Russell's diary is included here, unedited, as a snapshot of what the campaign was like on a daily basis for many protesters.

'Daily Update...
I started writing this down, the action around Sheffield's street trees, because it was so fast and furious that I needed to get each day's events clear in my mind. This is what I saw and wrote at the time.'

(Heather Russell, Tree Protester, February 2022).

News Round-up Wednesday 28th February 2018

SNOW...well you might have noticed.

There are some brilliant reports of what happened today on Thornsett Road.

Please look on this STAG FB page and Save Netheredge Trees FB Page and have a read.

Here is a very brief timeline this morning's events:

8:00	They came and they gritted, and gritted and gritted Thornsett Road.
8:15	Trucks left Olive Grove. We think they went for a long breakfast....
10.21	Amey Yellow Jacket Army arrives on Thornsett Road
10.22	Well wrapped up local residents and campaigners materialize through the light snow.
10:30	Darren Butt calls Social Services because kids were sledging on the pavement on a school 'snow closure' day.
10:38	Slow walk in front of Acorn Truck at top of Thornsett delays set up.

Police here.
11:00 Thornsett Rd resident was (allegedly) kicked by an Amey operative.
11.05 Setting up for felling stops. Amey operative leaves with police.

Everyone stands around for a while....much chatting and texting

11:16 Excited message from a resident "Great news! Just been informed by police no felling today on Thornsett"

Everyone stands around for a while longer....much chatting and texting

11:35 Report of Amey Vehicles leaving Thornsett
11:48 Darren Butt says no felling anywhere today.

Heavy snow now. Good call Darren.

11:51 Video on Twitter of PC Rick saying no felling in Sheffield today.
11:55 Tree Protester reports "Beast from the East defeats Amey tree felling on Thornsett!"

.....with a little help from tree friends....
Ta Da! Well done Tree Protectors!
So far on Thornsett, three days, two trees felled.
Snow and High winds forecast for tomorrow and Friday
P.S. Where can I join the Thornsett Road sledging club?

Round up in selected messages from various groups today:

8:15 Barrier trucks just left Olive Grove Depot. Acorn cherry picker in Asda car park Queens Rd. Please be alert on threatened roads
8:22 Gritter up and down Thornsett today. They don't have a lot of imagination
9:50 In addition to snow today, high winds forecast for tomorrow and Friday.

9:55	Works for me Just hope normal citizens don't get injured
10:01	They are wasting salt on felling healthy trees and not gritting roads that need it. Surely this is unsafe?
10:21	They have arrived Thornsett Agden
10:38	Slow walk Acorn Truck at top of Thornsett. Police here.
11:00	Resident Thornsett Rd A resident got kicked and was being incredibly aggressive. Protesters shouting "he's not with the protests"
11:15	Thornsett Slow walking to delay work
11:16	Just been informed by police no felling today on thornsett
11:17	Great news but look out everywhere!
11:20	From NE – Jason the arb just been on phone looking excited. Possible new location to go to?
11:25	Report of SIA being sent to another location from Thornsett.
11:25	King Edwards school has closed. Sending all who get in home. There will be kids around. Amey don't like that
11:35	Report of Amey Vehicles leaving Thornsett
11:45	Police say no felling on Thornsett today
11:55	Tree Protester reports "Beast from the East defeats Amey tree felling on Thornsett!"
12:00	Butt says no felling anywhere today. Heavy snow now

News Round up – Thursday March 1st

There was snow
No Street Trees were felled
Power to the Peaceful Snowballs

News Round up March 2nd 2018

The Good News:
    It didn't snow in Sheffield today
    Amey didn't fell any trees

The Not-So-Good News:

> Amey hard pruned the four Plane Trees in Fitzalan Square. When asked for a comment about the hard pruning an SCC Officer said: "The answer to your question about felling is yes. ...We needed to avoid birds starting to nest, hence the heavy pruning."

Petition you might like to sign

Petition to rescind the Streets Ahead Contract:

https://www.change.org/p/petition-to-rescind-the-streets-ahead...

Upcoming Events – click links for details

March 3rd   The Bentley Effect
Two screenings of "The Bentley Effect"
Saturday March 3rd 3:30pm and 6:30pm Meersbrook Hall
Tickets are still available.
March 8th   Meet the Chief – Sheffield
Thursday, March 8th at 6pm – 7:30pm.
The circle, Sheffield
Meet the Chief Constable
March 16th   Get Off Our Tree!
Friday 9:30pm STAG fundraiser "Get off our Tree"
The Everly Pregnant Brothers (Live) at Sheffield City Hall, Ballroom
March 22nd   Fundraising evening for Tales from Stump City Sheffield
                 The Gin bar Vintedge from 6pm

Have a great weekend.

Daily News Round Up – Monday March 5th

Abbeydale Park Rise – targeted today

> Early this morning the very first presence on APR was an Amey

vehicle, probably checking the condition of the road. Then there arrived a parking services car, tow truck and 4x4. A car was moved.

10.00 Saw the arrival of a massive convoy of vehicles; Barriers wagons, "welfare" mini bus for the Amey workers, chippers, lorries to collect the chips from the chippers, police vans and various Amey cars and vans. Out of this convoy poured more than 30 police and over 40 Amey Yellow Jackets; Arbs, SIA operatives, Evidence Gatherers, Barrier men, an Ecologist and assistant.

Amey barrier men set up Herras barriers at the bottom three trees. Then took down Herras barriers at two of them; the ecologist there said there was a birds nest in one tree and the other one opposite was too close to the nest to allow felling. This left one tree with Herras barriers around it and a Gecko in the hedge.

11:15 A Slow protest walk began and just happened to be in front of the Herras barrier van, and other vehicles, Amey wanted to move up the road. Eventually there were over 20 Tree Protectors having a lovely stroll back and forth across the road.

Meanwhile Acorn Arbs hand sawed and lopped the roadside canopy off the first tree they had set up around at the bottom of the road. While the Arbs sawed away two residents protected the tree where it oversailed their property by climbing up ladders and leisurely pruning the Holly beside the threatened tree.

12:30 Back at the slow walk...after a little over an hour, and about 10 metres moved, the police decided to form a thin yellow line and "kettle" all the strolling tree Protectors up and to one side of the road. Several people fell down and two ladies felt unwell and, after a rest on the road, moved out of the kettled area. Once the police had most people to the side of the road a charming Frenchman sat down and played his Guitar and sang. This prompted an outbreak of "Strictly Come Protesting" Ballroom Dancing, very elegant.

13:00 onwards... Barrier wagons were brought in from the top end of the road. After setting up around a tree that was not on any threatened list (we did tell them...) the crew finally managed to get barriers up at three more trees, all part felled trees they had previously attacked.

Geckos got in everywhere, Gnomes stood in garden permissions and a French squirrel stood triumphant in a tree while the admiring crowd sang their version of La *Marseillaise*.

Yorkshire Mix sweeties were offered to all; we are an equal opportunity Tree Protecting Protest group.

By the end of the afternoon:

No more work had been done on any of the trees. All day only tree branches were removed from the one tree at the bottom of the road.

One person had been arrested for not giving their name quickly enough and we think the policeman was impatient and wanted to practice his handcuff technique. Once arrested the person was de-arrested. They may get a summons for a 303 Highways offence.

A second person gave their details quickly enough when asked and avoided the handcuffs. They were told to expect a summary charge for breach of 303 of the highways act. Not a recordable offence, should they be found guilty.

A third person was arrested, not quite sure why, possibly impersonating a nesting French hen.

A woman inside an incomplete zone collapsed after being removed. She was eventually taken to hospital in a police van as the ambulance service said there would be 4 hour wait. We understand that she is recovering well in hospital.

Please look on the STAG FB page for some interesting videos, pictures and descriptions of today's events on APR and everything I've missed. It was a busy day...

POWER TO THE PEACEFUL.
WE ARE NOT GOING AWAY.

Daily News Round-up – Tuesday March 6th

This morning Yorkshire Water was fixing a leak on Abbeydale Park Rise and Amey's attentions returned to Thornsett Road, Netheredge.

It was almost déjà vu but at a different tree, with more singing, lots of Gnomes, more chanting, no arrests, no ballroom dancing and no trees felled – again.

The huge Herras "safety zone" completely blocked Thornsett Road. Surrounding and inside it were: One PC Rick (of impressive moustache fame), two Police Inspectors, three Police Sergeants, two Police Liaison Officers, two Police Evidence Gatherers, twenty three Police Constables, all backed up by numerous Police transport vans and a Police CCTV mobile surveillance centre with a camera mounted high on the biggest periscope I've ever seen.

Plus the usual compliment of around 40 Amey/Acorn Yellow jackets and, today, two Amey Senior Management. Senior Management were not totally keen to talk to the many representatives of the press, TV, Radio and a couple of journalism students who all arrived with impressive amounts of recording equipment. Hope we see some good reporting tonight and in the following days.

Megaphones at the ready there was some impressive chanting from a large crowd of Tree Protectors:

"Axe PFIs – Not Trees!"
"Whose Trees? – Our Trees!"
"Whose Police? Our police!"
"Plant Acorns – Don't employ them"
"Who lives here? – We live here!"

Please feel free to suggest new chants in the comments below.

Shortly after lunch time the Arbs had removed the roadside branches from one Chestnut tree that the Independent Tree Panel assessed as "Healthy". However, the Amey Tree Specialist assessed it today as "Diseased, rotting from within and needing to be felled within two years" Well, who to believe? I'll leave that up to you.....

There was Garden Permission beside the tree, which allowed the Gnomes to be under the oversailing branches. Amey were not willing to cut above the heads of the Tree Protectors and we all went home...rather slowly, strolling in front of some big vehicles...

Oh and the singing...we need more songs. We did have a wonderful rich baritone who led us in a short, but spirited, rendition of "He's a Lumberjack and he's OK". The Amey Lumberjacks were probably not overly pleased with the second line or the suggested clothes choices but we had a good sing.

So, polite peaceful homework for tonight: more songs and some creative chants.
POWER TO THE PEACEFULLY TUNEFUL
WE ARE NOT GOING AWAY.

Daily News Round-up – Wednesday March 7th

Today was a good day, no trees felled.

Amey and Acorn dug holes, planted saplings, moved logs, ground stumps, worked on lighting, cleaned pavements, patched footpaths, did good stuff.

Thank you.
We like this.
Have a lovely evening.

Daily News Round-up – March 8th 2018

It's been a long day for a lot of people
Starting messages:

5:10   "Look out of the window. Snowing a lot in S2"

7:18   "Good morning you lovely lot. Welfare van has left the depot. Heras Barrier vans still in yard and other vehicles"

Finishing messages:

19:01   "Two trees down. 45 minute hold up after they were ready to go home. Quite a few strollers out for an evening constitutional across the road"

A whole lot happened in between.

All morning, there were cat and mouse games with barrier vans and Security vans coming and going.

Is it Abbeydale Park Rise or Kenwood Road?

Then it all kicked off at Kenwood Road at 12:00.

We had the usual complement of Police, SIA Security and Arbs plus the Police and Crime Commissioner Dr Alan Billings and lots of

Tree Protectors.

There were barriers, Barrier vans, welfare vans, bunnies under vehicles, Cherry pickers, songs, chants, sweeties handed round, politicians, reporters, BBC, Gnomes in Garden Permissions, Gnome in a holly bush, Dog walkers, mothers with babies, school kids walking home after 3pm, people needing to visit the surrounded property and…anger, anguish and… tears.

Tears as everything we did could not prevent the felling of two fine, treasured, valuable trees.

Please have a look at the accounts and video footage on STAG FB page, they tell it all.

POWER TO THE PEACEFUL
WE ARE NOT GOING AWAY

Daily News Round-up – Friday March 9th 2018

It occurs to me that many people associated with Sheffield's Tree campaign will feel as I do today. Really glad it's Friday but also exhausted, upset, disillusioned, angry, betrayed.

Today one Cherry tree, already missing branches, was felled on Abbeydale Park Rise.

This required:
One Acorn Arborist crew of two.
Nine police vans, each seat taken by a police officer.
Enough Private Security SIAs to control Glastonbury festival.
This Cherry makes the third tree felled this week.
Estimated cost – £20,000 per tree.
That, in itself, is enough to make anyone feel depressed, worse was

to come

Once the tree was felled, and everyone was going home quietly, two arrests were made.

Both arrests relied on, what I think are, spurious accusations made by Private Security SIAs.

The Police are obliged to investigate.
There was also a notice of summons on two counts for a third person.

Amey and Sheffield City council have made a deal with the devil bringing in Private Security to oppress citizens.

POWER TO THE PEACEFUL
Have a restful weekend
We have work to do next week.

Daily Round up – Monday March 12th 2018

On the streets it was a quiet rainy day; nobody out and about felling, just a few stumps getting ground and a few pot holes filled.

The fireworks were in the Local and National Press as Sheffield City Council have been forced to reveal the redacted parts of their contract with Amey that deal with felling of our Street Trees. There is a target to fell 17,500 street trees.

The much vaunted engineering solutions are a fiction.

Good reads in the Times and *Daily Mail*. Look out for the *Star*, *Yorkshire Post* and our own Jennifer Saul in Medium.

If there are any fans in the Sheffield City Council offices they will need a good clean after today.

Daily Round up – Tuesday March 13th 2018

Unlucky thirteen for the beautiful tree outside the Bridge Club on Thornsett Road, Netheredge.

> Despite having members who support the tree campaign, including the Rustlings Road two, the Bridge Club Committee decided to deny Garden Permission, which could possibly have saved the tree.
>
> One person was arrested while peacefully defending the tree. Over 50 Tree Defenders were spread out around the very large workzone and joined in chanting led by two people with heroically loud voices.
>
> The attack started shortly after 10am and finished at 5:45pm with tree protesters slow walking in front of the convoy of Acorn trucks and herras barriers as they left. There were fewer police in attendance today, 12 at mid day, plus the CCTV van.

Do listen on catch up to Brian Lodge on this mornings' Toby Foster at Breakfast – BBC Radio Sheffield. I love a comedy turn in the morning.

Eyes on the roads all areas tomorrow.
POWER TO THE PEACEFUL
We are not going away.

News Round-up – Wednesday March 14th

Wednesday and no trees have been felled, just like last Wednesday. Good. More good news, please read Louise Haigh in The Star: https://www.thestar.co.uk/news/sheffield-mp-louise-haigh-calls-again-for-immediate-halt-of-tree-felling-works-and-blasts-city-s-highways-pfi-contract-1-9062556

Have a lovely evening.
Keep watching tomorrow, like Arnie "they'll be back"
POWER TO THE PEACEFUL
WE ARE NOT GOING AWAY

Daily Round up – March 15th 2018

Kenwood Road is where it all happened today. There was the usual compliment of Security, Barrier men and Arbs but there appeared to be fewer Police actually on the street, however there were a couple of Police vans parked a little way away with blacked out windows.....

> Amey seem to have found the correct metal clips to put together the Heras barriers rather than the cable ties they usually use. I wonder if it has anything to do with the use of cable ties invalidating the insurance when Heras barriers are tied with them instead of the solid metal clips.?
>
> Acorn spent a long time not felling any trees as everyone watched. We had garden permissions and a lot of Tree Protectors. The Netheredge Slow Walking Society held a practice shuffle and there was some lovely singing as the vehicles, very slowly, left. One Tree Protector was invited to travel with the Police to Shepcote Lane. We hope they will be home soon.
>
> So... no trees felled today and all three Sheffield Labour MPs call for mediation over the Tree fellings.

Despite the rain all in all a pretty good day.
POWER TO THE PEACEFUL
WEARE NOT GOING AWAY

Daily News Round-up – March 16th 2018

Wow, what a day of waiting, highs, lows, Samba, Jarvis Cocker, BBC Radio Four, BBC Two radio and, sadly, the loss of a Healthy Lime Tree.

> Quote of the Day: Jarvis Cocker on Radio 4 early this morning,
> "...There are the 6 D's for the Sheffield Trees now there's a 7th ... DAFT"

7:00   Olive Grove lots of checks done on vehicles

Then yawn for over three hours while we all held our breath... metaphorically...until,

10:15   All in Bramall Lane Petrol Station taking it in turns to go into Subway (new Police Canteen?)
There were:
8 Police riot vans,
2 normal Police cars,
2 Enterprise Rental vans with police inside them,
3 unmarked cars
1 Amey car
Then Servoca Private Security (the yellow jackets) joined them. Too much DayGlo HiViz...hope the Petrol Station had sunglasses on special offer this morning.

10:25   Convoy spotted at Brook Hill Roundabout

10:39   Heras Barriers go up round tree at 58 Sackville Road, Crookes
Despite the Independent Tree Panel recommending this healthy tree should be saved by simply putting in a half kerb, SCC went ahead and approved felling. Does the 17,500 contractual obligation have anything to do with this? Also interesting that a person at Sackville Road reports that the road and pavement have already been "upgraded"

11:16 Gennel allowing Public access to Sackville Road completely blocked by Heras Barriers and Servoca Yellow Jackets. We have the pictures.

11:35 Sheffield University Samba Band arrives, Rhythm, dancing, grinning and high spirits break out and beat back the sound of the chainsaws. Thank you Samba Band!

12.00 BBC Radio Two Jeremy Vine discusses Sheffield's Street Trees. Do listen on playback.
Paul Brooke spoke brilliantly, despite the connection being lost once. Brian Lodge also spoke.
JV said they had about a trees' worth of texts, tweets and comments about our trees.
We could hear the samba band in the back ground, and the chainsaws, as a reporter on Sackville Road described the scene as #58 was destroyed.

13:25 Goodbye healthy Lime at #58 Sackville Road, Crookes, now just a stump and a pile of logs.
The other targeted tree on Sackville Road has a bird's nest in it and so cannot be touched
Good timing birds, have fun looking after those eggs.

14:08 Peaceful Tree Protectors do the Sheffield Shuffle s-l-o-w-l-y up the road for 10 minutes, allowed by the police.

22:00 Those lucky enough to get tickets, See you at the Sheffield City Hall Tree Gig.
7th D....DAFT – Thanks Jarvis

POWER TO THE PEACEFUL
WE ARE NOT GOING AWAY
and possibly not sleeping tonight...

Daily Round up – Monday March 19th

Very quiet Tree day today, not complaining though as no Street Trees were felled in Sheffield.

>Some stump grinding was done.
>Crew with a stump grinder turned up to an almost complete Cherry tree.
>Looked at it for a while and left.
>Came back to be met by some residents.
>Went away again.
>Lone Arb bought lunch at the station.
>Yes, it really has been that quiet.

Might be a busy day tomorrow.

POWER TO THE PEACEFUL
WE ARE NOT GOING AWAY

Daily Round Up – Tuesday March 20th

Today Rivelin Valley Road, the second longest Lime avenue in England, lost TWO of its majestic, healthy, over one-hundred-year-old Lime trees. This avenue is situated in Paul Blomfield's constituency who, only a week ago, said as part of a long statement:

> "I would also like to see a pause in the work for more discussion to resolve the current conflict, putting the views of residents on affected streets first." It looks like that would be a "no" from SCC and AMEY

> Rivelin Valley was invaded this morning with the now customary resource wasting complement of large numbers of Police, Servoca Security employees, Amey barrier people and employees of **Acorn Environmental Management Group** – who actually cut

the trees down.

By the end of the day this invasion resulted in two trees being felled and two people, who came to protest peacefully, being arrested.

Just to give you the feel of today here is a selection of comments posted on FB and various chats:

"SIA. Heras.Police on RVR"
"felling?"
"think they're going for three trees"
"I'm at work...NO! NO! NO!"
"Have alerted Local Councillors"
"Live feed from Rivelin Valley Road shows Jeremy Willis really not looking pleased. He must have missed the memo about today being 'International Day of Happiness'. "
On a Banner.."HONK FOR TREES"
"HONK!!!!!" From the many motorists who drove past showing their support
"South Yorkshire Police
PROTECTING the Fence
PROTECTING the Bouncers
PROTECTING the Contractors FROM
Peaceful People who just want to... PROTECT Healthy Trees"
"100 years to grow, and destroyed in 1 hour"
"Shameful butchery"
"Damn – they are ruining one of Britain's best beauty spots."
"A sad, difficult day."

"15:45 All finished at Rivelin Valley Rd, Thanks everyone who came"

I think that means thanks to the Tree Protectors....

POWER TO THE PEACEFUL
WE ARE NOT GOING AWAY

Daily Round up – March 21ˢᵗ 2018

And… they were back on Rivelin Valley Road this morning.

> 40 people in HiViz counted early on….Police, Acorn, Amey Barrier men and Servoca Security Staff.

Let's just pause there for a moment to think a bit about the Lime trees all along Rivelin Valley Road.

> Way back in 1906 over 700 Lime Trees were purchased from Dixon's of Chester at a cost of £147.

> The Limes were planted to line the road for a distance of 3.5 miles, making it the second longest Lime Tree Avenue in Britain, and it still is today.

> These trees are a Sheffield community asset. One 112 year old Lime tree has a CAVAT (monetary amenity asset value) of about £30,000 pounds. Almost 600 out of the original 700 trees are still in place and healthy.

> The total the CAVAT value of the Avenue, at a conservative estimate, is: £30,000 X 600 = £18,000,000
> 18 Million Pounds…..hold that thought.

So.. back to this morning at Rivelin Valley Road….

> 40 people in HiViz arrived to facilitate the felling of targeted trees in the Avenue. They were rapidly joined by many local residents and some Tree Protectors from other parts of Sheffield.

> 7 or 8 geckos got behind the targeted tree and prevented the completion of a barrier zone.

This held up any planned felling for a long time.

Three warnings were issued then security staff used considerable force to remove the Geckos.

During the forcible Gecko removal a female protester (age 61) was dragged to ground by security staff. She was then kicked in the head by one of them as he stepped over her (there was plenty of room to walk around her).

She was taken to A&E by ambulance and was found to be suffering from a severe headache caused by a big bump on her head. She also had a badly bruised hip from where she hit the floor when she was dropped.

Police filmed the entire incident but refused to take a complaint from the victim at the scene. They said she would have to go to a police station to make her complaint. The difference; compared to immediate arrest if the security personnel complain; is astonishing.

Following this:
- Another protester got back into Gecko position and was eventually arrested and taken away by police.
- A local resident was arrested for swearing and taken away by police
- A lady, who had been enthusiastically blowing a Zuzuzuala most of the time, was arrested for, well, blowing her Zuzuzuala. When the Zuzuzuala was taken away from her she set off her rape alarm. The rape alarm was also taken away from her and she was taken away by police.

Surrounded by Heras barriers, security and police the Acorn Arbs felled the remains of one, already partly felled, healthy Lime tree. They all left shortly after one thirty.

So by lunchtime; one lady had been injured and was in hospital; three people had been arrested and were awaiting processing in the quaintly named Snig Hill Custody Suite; one healthy Lime tree had been felled.

1That makes it 30.000 pounds off the value of the 18million pound Rivelin Valley Road Lime Tree Avenue.
And how much spent on police, security and support services?
What a terrible waste.

POWER TO THE PEACEFUL.
TIME TO STOP AND TALK
WE ARE NOT GOING AWAY

Anything can happen in leafy Norton, before coffee

Thorpe House Rise – Norton S8, early Summer of 2017. These were still quite gentle days in the Tree Campaign when there was no injunction, no security guards and low plastic barriers were used around trees when a felling was attempted.

Before my first coffee one morning I saw an alert on Facebook that an Amey chipper and an Arb van had been seen heading for the Norton area. That is very near where I live so I walked up to have a look. From some roads away I heard the distinctive noise of a chain saw. My heart rate sky rocketed as I rounded the corner and saw a barriered off tree with an Arb already up it lopping off branches. I yelled to him that I was going into the work zone to stop the felling. All work stopped and as the Arb climbed down I went in over the plastic barriers and sat under the tree.

At this point an elderly lady with bright orange hair came out of one of the houses opposite.
She started berating me with familiar challenges...

"Haven't you got anything better to do?"
"This is no business of yours you don't come from round here"
...and so on but then she got a bit more creative.
Calling over the Arb she pointed to one of the higher branches directly above my head and said,
"See that branch there?"
The Arb agreed that he could see the branch. She continued with considerable enthusiasm,
"Well, I want you to go up there and cut down that branch and drop it on her head and KILL HER!"
The Arb, somewhat surprised, said "I can't do that Madam"
This reply did not please her at all and she started shouting,
"If you won't do it then I'll KILL HER! That'll stop her I'll KILL HER! I will I'll KILL HER!"
She continued making death threats from the other side of the barrier for quite a while. At this point I found my phone was out of charge, shame I wish I had a recording.
The Arb came and stood beside me for a while and I told him I would not report the death threats to the police as I didn't think the woman was entirely rational. I wish now I had made a report.

The tree survived that day.

Lessons learned:
    Don't forget to charge your phone
    Always report death threats
    Anything can happen in leafy Norton, before coffee.

Assault by a resident on a Tree protester – resulting in a Police Caution for the resident.

The assault occurred on May 24th 2017 at 8:50am, outside number 23/25 Thorpe House Rise S8 9NP.
The tree protester assaulted was Heather Russell.

The Thorpe House Rise resident who committed the assault was Mick Mathers.
Police Incident number 2212240517.
Officer attending the 999 call was P.C. Oades, Woodseats Police Station.

Report by Heather Russell:
On the morning of May 24th 2017 I saw a message to say that there were multiple felling crews setting up to fell trees on Thorpe House Rise.
I walked to Thorpe House Rise and when I arrived I found that protesters had entered the work zones and stopped the felling of all of the threatened trees except the tree outside number 23/25.
Paul Brooke was in the work zone of the first tree I came to. He said " We have all these tree covered but they are still cutting one further up the road. See what you can do about that"
I walked up the road and alerted the crew felling the tree at 23/25 that I intended to enter the work zone and stop the felling.
A Sheffield City Council employed person began filming me as I entered the work zone by squeezing in between the hedge outside number 25 and the plastic barrier.
As I went in I became aware of loud shouting and a man leapt over the front barrier and ran at me.
He grabbed me by the shoulders and pushed me into the hedge, then with one hand slightly moved the barrier.
Shouting all the while he then grabbed me with both hands and threw me out of the work zone past and over the plastic barrier.
I just managed to avoid falling flat on my face.
I was by this time shouting repeatedly "I am being assaulted call 999, call the police"
He was shouting and swearing very aggressively in my face the entire time.
The SSC employee continued to film the entire assault.

I called 999 with the help of another protester who had arrived as I

was shaking so hard I could not operate my mobile phone.
The Police arrived and took both myself and the person who had assaulted me to Woodseats Police Station in different cars.

After I gave my statement to the police I was asked which of three outcomes I wanted for my attacker to face:
1) Restorative justice
2) Police Caution
3) Charged and taken to Court

I said I wanted the incident to go to Court.
I was then told that the Officer in charge would make the decision between offering my attacker a Police Caution or being charged and taken to Court.
I was told what was decided depended on whether the person had a Police record or not.

My attacker, Mick Mathers, initially denied the attack but was shown the film of his actions.
He accepted that he committed the assault and was given a Police Caution.
This all was completed the same day, within a few hours of the assault.

Two days later on the morning of Friday May 26th Mick Mathers shouted at me from his white van and tried to intimidate me while I was on Thorpe House Rise with a friend.
I reported this intimidation to the police via 101 #87326/05/2017.
Nothing was done about it as far as I know.

I feel that Mick Mathers was treated in a markedly different manner from the way Tree protesters accused of assault on Jan 22nd 2018 on Meersbrook Park Road have been, and continue to be treated.
They have been arrested, held for hours in cells, bailed, and re-bailed twice and one has been arrested and accused of intimidation which was then dropped.

They have now been called to Shepcote Lane Police Station for the evening of June 4th.
They have not been given any indication of what will happen on that evening.
The whole affair has been drawn out over four months.
The Police have not been even handed and impartial in their handling of these assaults.

STANLEY ROAD, S8

The first set of Security Guards, the ones recruited from the job centre and sent on a one week bouncer course, tended to have a short fuse.

Stanley Road, Summer 2017

Lone Bunny enters barriered work zone to prevent a felling.
Security Guard tries to get the six man Arb crew to surround the Bunny. They refuse.
The 6' 2" Security guard, in full Hi Viz suit and helmet, then strides up to 5'5" lone 68 year old Bunny until his chest touches her nose and robotically says
"You – are – in – my- personal – space. I – feel – intimidated"
Bunny bursts out laughing.
Security guard gets mad and picks up a plastic barrier and slams it down on Bunnies foot.

Bunny reports incident to the Police.
Security Guard looses job.

I encountered the same Sheffield AMEY Arb crew the following day while stopping them felling on Lismore Road. They apologised for the behavior of the Security Guard the previous day.

Lismore Road

In the very early days of Heras barriers
Two crews descended on Lismore Road intent on taking as many of the nine threatened trees as possible. They arrived with the new high metal Heras barriers.
Geckos clung to walls, Gnomes stood in gardens and Bunnies tried to get in. The campaigns slowed the work considerably but one tree was taken that day.
When it came to packing up the barrier crew decided to leave all the barriers in a big pile on Lismore Road rather than load them on a truck and take them back to Olive Grove. They decided it would save them time in the morning.
A few campaigners stood round regarding the Herras pile. Someone, a keen cyclist, remarked that a few cycle locks could make the pile hard to move.
Word spread quickly. People searched sheds, dug our bike chains from the back of drawers and took them off their kids bikes.
B&Q had a selling run on solid cycle locks, bike chains and all varieties of padlocks. People ran into each other buying lock up equipment in hardware stores.
The chain up started in day light, carried on after dark and right through the night.
One campaigner visited several times, in feminine disguise [Simon Crump], with several locks each time.
By morning the Harras Barrier pile was festooned and tangled with a multitude of locks.
They looked very festive.
The barrier crew were not best pleased when they arrived in the morning. Clippers were produced, and promptly broke attempting to take the locks off. Seems bike looks are not supposed to be easy to remove. Who knew?
They sent off for bigger clippers (which broke) and an angle grinder. It transpired that the angle grinder could not be used as they couldn't clamp the locks and the possibility of bits flying off

was a Health & Safety hazard.

The day wore on and more types of cutters were tried without success. Slow removal progress finally happened when cutters resembling the 'Jaws of Life' arrived.

The barriers could finally be moved by packing up time. The barrier crews never left their barriers behind in a pile again. Shame, that was fun.

ACKNOWLEDGEMENTS

We are very grateful to all the interviewees and contributors to the story. They lived it all, took part in the actions, and then re-lived their experiences for this book. Every one of them has played their part in important events.

We are also grateful to those supporters who donated towards the costs of producing this work. We hope that the finished book is our thank you.

We would like to thank the following people for their support and advice:

David Alcock
Luis Arroyo
Paul Brooke
Sheldon Hall
Rebecca Hammond
Nick Hayes
Russell Johnson
Helen Kemp
Christine King
Jane Miller
Mary Ren
Paul Selby
Gary Stimson